J H Rowland

FARM MACHINERY

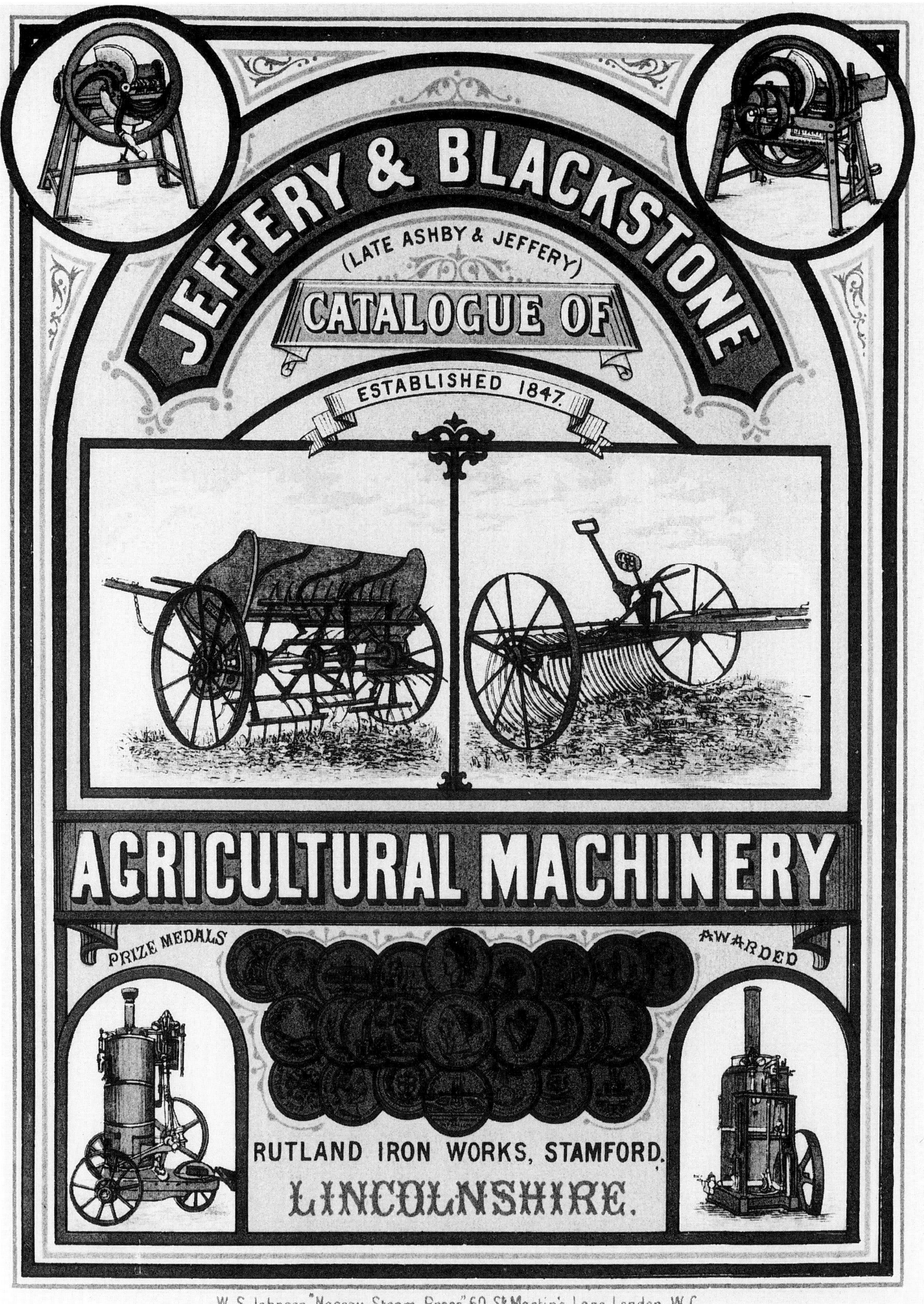

Catalogue of agricultural machinery by Jeffery & Blackstone of Stamford, 1884. [35/26382]

FARM MACHINERY
1750—1945

JONATHAN BROWN

B.T. Batsford Ltd, London

First published 1989

ISBN 0 7134 6100 4

Typeset by Servis Filmsetting Ltd, Manchester
printed in Great Britain by
Mackays of Chatham plc, Letchworth
for the publishers
B.T. Batsford Ltd
4 Fitzhardinge Street
London W1H 0AH

Contents

Illustrations

The illustrations in this book are all drawn from the collections of the Institute of Agricultural History & Museum of English Rural Life, University of Reading, including material from the photographic collections of Eric Guy and *Farmer & Stockbreeder* held by the Institute. [Catalogue numbers are in square brackets.]

ONE

The speed with which farming has become a thoroughly mechanized industry since the Second World War has become almost proverbial. In no more than three decades working horses disappeared, manpower was drastically reduced and the land worked instead by the mechanical power of tractors, combine harvesters, forage harvesters, precision drills, tractor-mounted sprays, and a host of other implements. Rapid and complete though this process of mechanization has been, it was built on a firm foundation of development of agricultural implements extending over a period of 200 years. It is the course of that development, from the improvements to ploughs and the first successful seed drills of the early eighteenth century to the first generations of combine harvesters to work in this country, that is the theme of this book.

The theme is treated in four broad chronological sections, for each of which the text points to the main features of development in farm machinery: the eighteenth-century improvements in ploughs; the growth of a new agricultural engineering industry between about 1790 and 1850; agriculture's steam age in the second half of the nineteenth century; the new power of the internal combustion engine in the twentieth century.

As well as these main developments there were numerous other inventions and improvements to farm implements in each age, and as many as possible of these are included in the illustrations and commentary for each chapter. Not all can be included, nor can many of the agricultural engineering firms be fully represented; that is inevitable with a volume of this length. Together, though, the illustrations and commentary provide a broad survey of the process of mechanization in farming, of the implements in use on the farm, and a few of the novelties that never caught on.

1 *The several sorts of plough in England, illustrated in the* Universal Magazine *in 1748. Among those depicted are the Lincolnshire plough (a) with a wheel coulter thought better suited to the sedgy soils of the fens; the Hertfordshire wheel plough (h); the two-wheeled plough of Berkshire, Oxfordshire and Wiltshire (l); a four-coulter plough (d and k). Also shown is the seed drill of William Ellis of Little Gaddesden, Hertfordshire (c), a simply constructed barrow with rudimentary feed to deposit the seed in the furrow. [35/10318]*

2 *Ploughing with tractor power, Wiltshire, May 1940.*
[35/26385]

TWO
The Eighteenth Century: farm implements before industrialization

When eighteenth-century writers on agriculture such as William Marshall and Arthur Young turned their attention to farm implements they filled their pages almost entirely with a consideration of the types of plough then in use, and the possibilities for their improvement. They had good reason to do so for the plough was then the most substantial of the tools of tillage. It was one of the most expensive pieces of equipment on the farm. The common Devonshire plough, relatively simple in construction and usually made by a local hedgerow carpenter, cost about 15 shillings to buy new, at a time when a simple, wooden-framed harrow cost a shilling or two. Besides the plough, farm implements were simple: either hand tools, such as the hoe and the flail, or horse-drawn implements such as the harrow and the roller.

What most concerned the authorities on agricultural affairs when they turned to the plough was its draught, its resistance to the soil, and therefore the horsepower needed to pull it. 'Every circumstance that lessens the expense of tillage, without lessening its efficacy, is of the first consideration in husbandry', William Marshall wrote as he introduced the subject of the ploughs used in the Midlands. The common types of the region, he thought, failed this test, for they were unusually heavy, clumsy and inefficient, requiring five or six horses to pull them.

There was a great variety of ploughs in use in the eighteenth century; it would seem almost as many different sorts as there were counties in England, and each adapted, in theory at least, to the peculiarities of the soil of its region. Some had long mouldboards, others short; some had shares roughly triangular in shape while on others the share came to a long, spiky point; some had a long beam, others quite a short one. The varieties were many, though often only of detail, and in broader terms all ploughs fell within a few general types which formed the basis for plough design throughout the era of horse power. There were wheel ploughs, which had one or, usually, two small wheels at the leading end of the beam. They were more usual in regions of heavy soil. A variant was the gallows plough, which had a steeply inclined beam supported by medium-sized wheels. The second type was the swing plough, which had no leading wheels, and tended to be the type used in light-soiled areas. Third was the turnwrest plough, a peculiarity of Kent, Surrey and Wealden Sussex. Its mouldboard and coulter could be moved from one side of the beam to the other, so that, as the plough followed its course up and down the field, all the furrows faced the same way across the field. The plough with fixed mouldboard turned all furrows to the right of the plough, which meant that on the return passage the furrow was turned to face that made on the outward course, and the field in consequence had to be ploughed in a series of ridges and furrows. The principle of the turnwrest plough remains with modern one-way ploughs drawn by tractors.

Arthur Young thought the common plough of the fens to be 'a most excellent tool', but such praise for the traditional ploughs was rare. More typical was William Marshall's judgement on the Kentish turnwrest plough. While acknowledging that it seemed well-suited to the flinty soils and steep hillsides of the downs, in almost all other respects he thought it to be the most unwieldy of all ploughs. The Hertfordshire wheel plough was held locally in high regard; notwithstanding that, Arthur Young held it up as an example of a plough that was 'heavy, ill-formed and ill-going', useful only for breaking up fallows.

Change, however, was well on the way. William Marshall noted with approval the spread of what he called the Yorkshire plough into the Midland counties, for these were lighter, required no more than three horses to pull them, and were easier for the ploughman to handle. These were almost certainly ploughs of the Rotherham type, the most influential new plough of the eighteenth century. Patents for a new design of plough had been taken out in 1730 by Joseph Foljambe of Sheffield. His main innovation was the use of a rigid triangular frame to which the share, breast and mouldboard were attached in place of the traditional practice of having a separate share beam to carry the share, which had been one of the main contributors to the unwieldiness of the old ploughs. The result was a plough of lighter weight and reduced draught on which the share, breast and mouldboard worked more smoothly together.

The Rotherham plough established the lines along which ploughs were to be developed for the remainder of the eighteenth and well into the nineteenth century. The size and shape of the mouldboard were subject to detailed investigation in efforts to find the most efficient form and lightest draught. The Scotsman James Small published a *Treatise of Ploughs and Wheel Carriages* in 1784 and designed an improved plough that became quite popular in Scotland. James Arbuthnot, a Norfolk farmer, in the 1770s made mathematical calculations as to the best shape of mouldboard, and produced a swing plough intended for two-horse traction that enormously impressed Arthur Young. Young was also impressed by the work of another East Anglian, John Brand of Lawford in Essex, who made a plough entirely of iron that became known as the Suffolk iron plough. It had only one handle, a novelty that continued on ploughs of Suffolk type into the 1840s. Lord Somerville, one of the landowners promoting agricultural improvement at this time, made his own improved version of the two-furrow plough used in the southwest.

Despite all this inventive activity the pace of change was relatively slow, as the complaints about the general standard of ploughs in the 1780s and 1790s show. The uptake of new implements depended on the willingness of farmers to see that there might be advantages in using them, and also upon their being readily available. Joseph Foljambe had shown the way to make the implements available in establishing a factory to produce his ploughs in Rotherham (whence their name was derived). Here he had the capacity to produce about 300 ploughs a year using standard wooden parts derived from master patterns. Foljambe's factory, however, remained a novelty in the eighteenth century for large-scale production of agricultural implements did not become established until the nineteenth century.

The unwillingness of farmers to accept new implements was a major part of the story of one of the eighteenth century's most celebrated agricultural innovations: Jethro Tull's seed drill, devised in the early years of the eighteenth century. He was not the first to seek mechanical means of sowing seed, for several inventive minds in this country and on the continent had, since the sixteenth century, been applied to this problem. Tull's invention stands out from the others for several reasons. In the first place his drill evidently worked, and was used for a number of years on Tull's own farm, while the inventions of others generally came to nothing. A second prominent attribute was the amount of publicity he achieved, mainly through his book *The Horse Hoeing Husbandry*, first published in 1731 and enlarged in 1733.

Despite the successful operation of the drill and the publicity surrounding Tull, when Arthur Young was examining the state of English farming 60 years later, very few farmers were using seed drills. That was because for most farmers there were neither technical nor commercial advantages to be gained from the drill. Mechanically, seed drills were still inefficient, often unable to deal with stony or uneven ground, liable to become clogged with earth, and therefore to sow the seed unevenly. They were expensive both to buy and maintain, and that meant returns on the crop had to be high to make investment in the machine worthwhile. For few farmers of the 1790s were those conditions fulfilled. As with the improved ploughs, the advance of the seed drill depended on the changes in the methods of farming and of manufacturing which were not to become general until the next century.

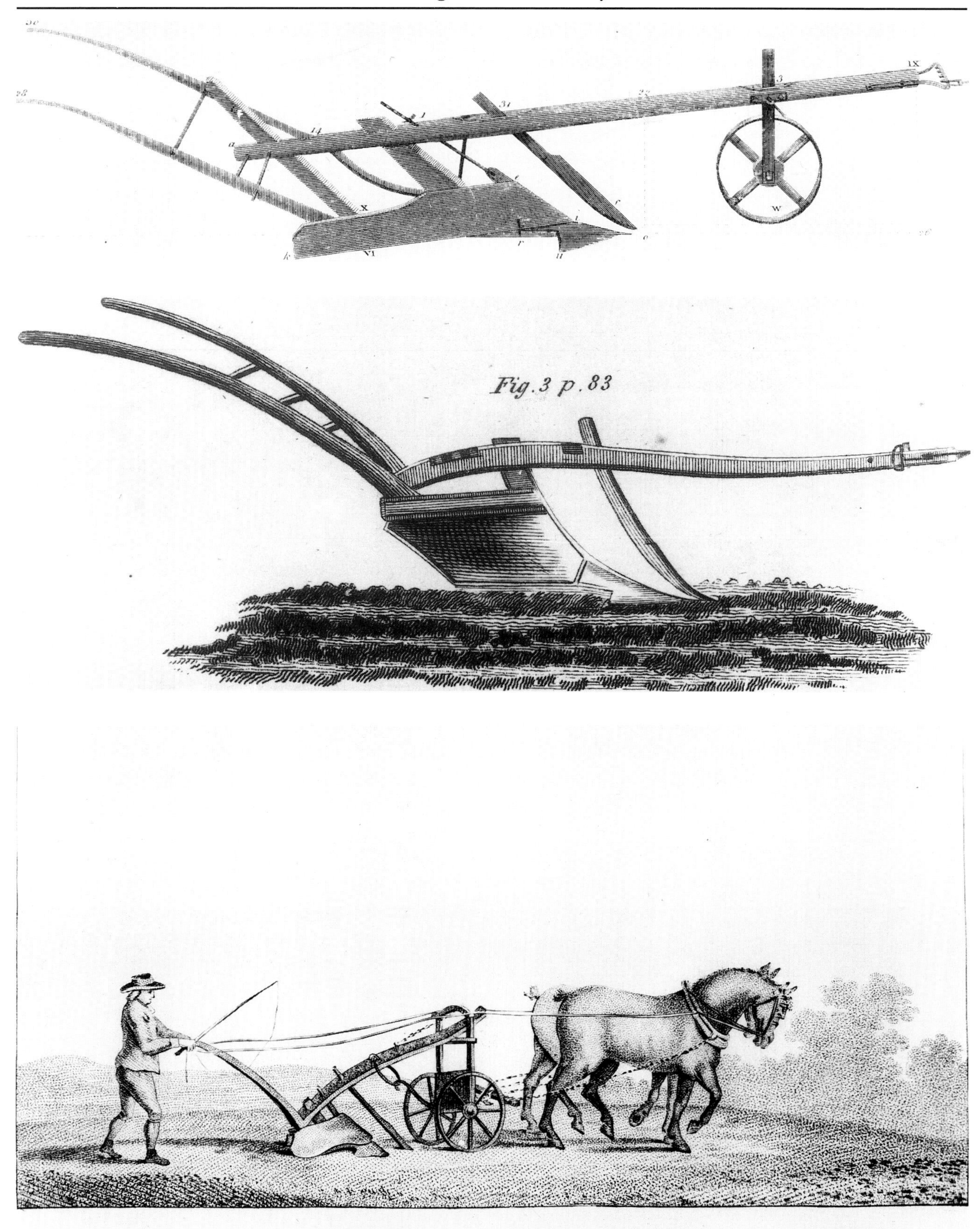

THE NORFOLK PLOUGH AT WORK.

Top left **3** *The old Gloucestershire plough, a single-wheel version which Arthur Young found in use in parts of Oxfordshire. The long beam and long, heavy mouldboard were characteristics of this plough, intended for use with heavy soils. [35/26323]*

Centre left **4** *A swing plough from the North Riding of Yorkshire. [35/1134]*

Bottom left **5** *The Norfolk gallows plough, illustrated in Nathaniel Kent,* General View of the Agriculture of Norfolk *(1796). [35/4323]*

6 *The Kentish turnwrest plough, shown at work in this illustration from John Boys,* General View of the Agriculture of Kent *(1805) and a restored example. The heaviness of the plough is indicated by the four horses needed to pull it, but Boys was certain that no plough was better suited to Wealden conditions. [35/8157 & 60/1852]*

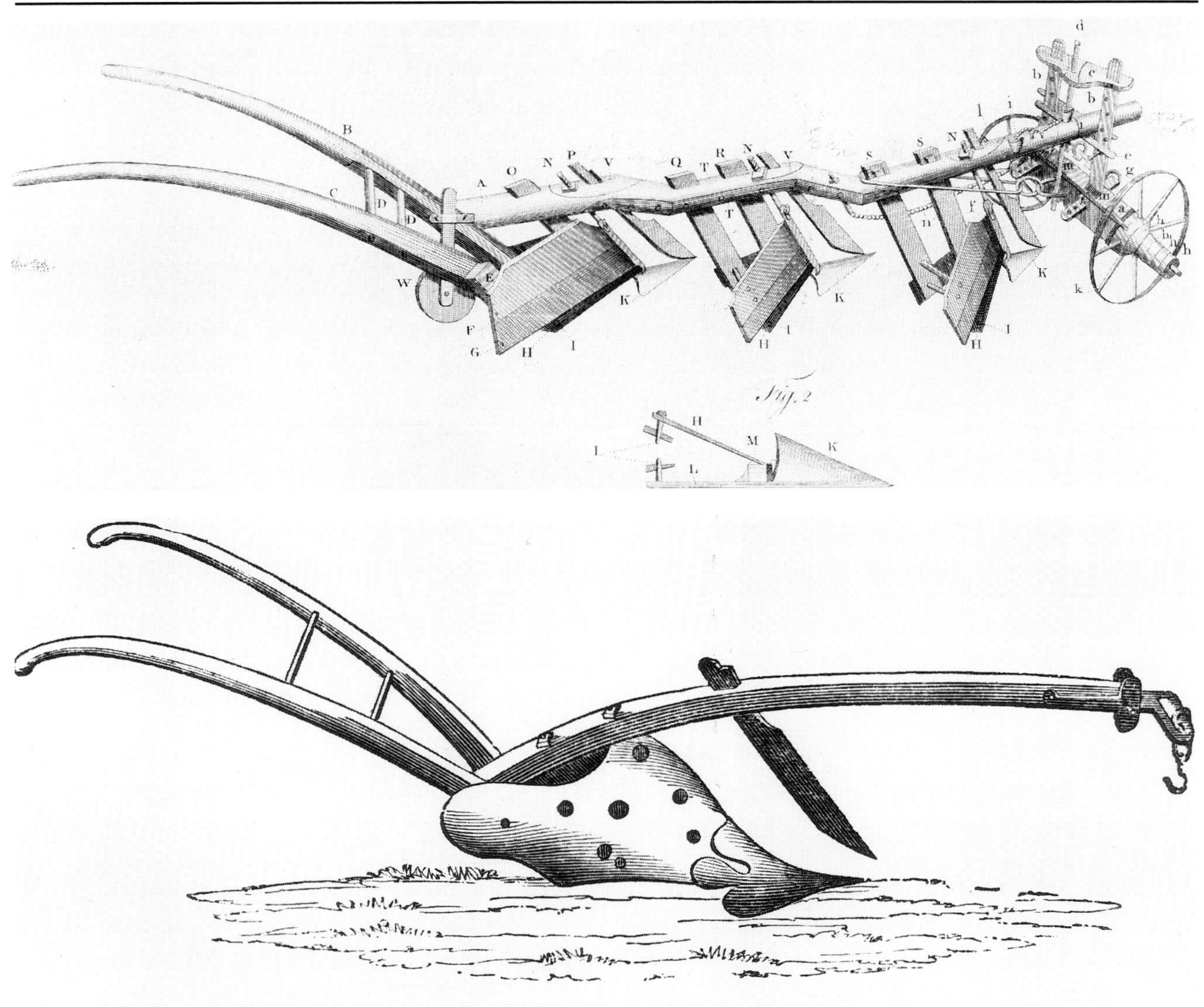

Top **7** *Double furrow ploughs had been in use since the seventeenth century at least, and during the eighteenth century various people brought forward improved designs. Mr. Duckett, of Esher, Surrey, achieved some repute, and his three-furrow plough was used by farmers in many parts of the country. The plough was introduced in the 1770s, and had a long, crooked beam supported by common wheels similar to the gallows plough. [35/11798]*

8 *The Rotherham plough, an illustration from J. Allen Ransome,* The Implements of Agriculture *(1843). Comparing this with the ploughs depicted in Illustration 1, Chapter 1 shows how the simplified frame construction of the Rotherham was producing a plough more elegant as well as less cumbersome. [35/17075]*

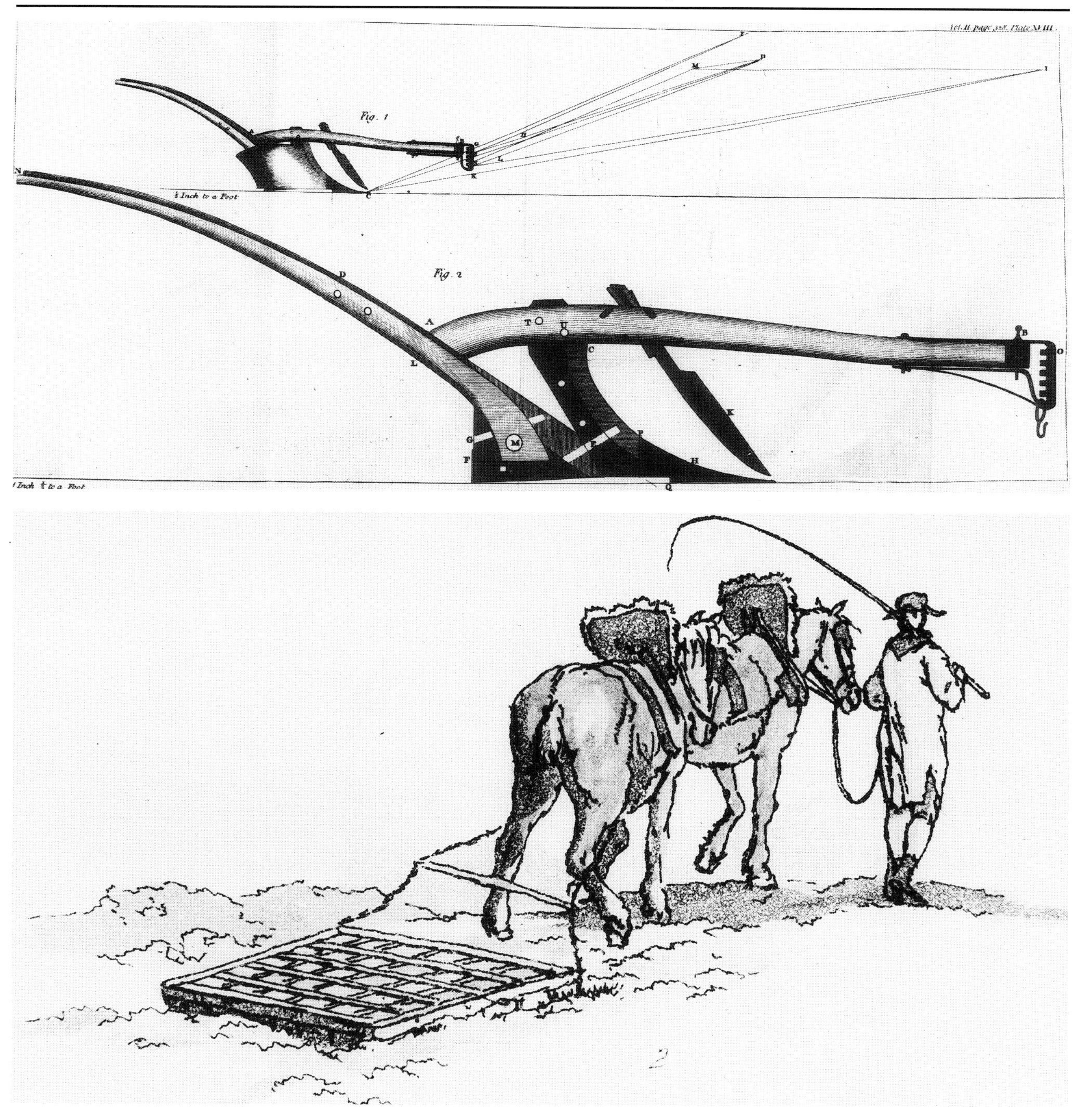

Top **9** *Arbuthnot's plough illustrated in Arthur Young's* Farmer's Tour Through the East of England *(1771). This was one of a number of diagrams intended to show the application of scientific methods to the design and construction of ploughs, and from which wheelwrights could learn how to make successful ploughs. [35/5279]*

10 *The most basic form of harrow, still in common use in the eighteenth century, was the bush harrow. It was quite simply a large bushy branch weighted down by a log or stone. Manufactured harrows of the period were usually of this pattern, a wooden frame—square, rhomboidal or triangular—with iron tines. W. H. Pyne's illustration from his* Microcosm *published in 1806. [60/14170]*

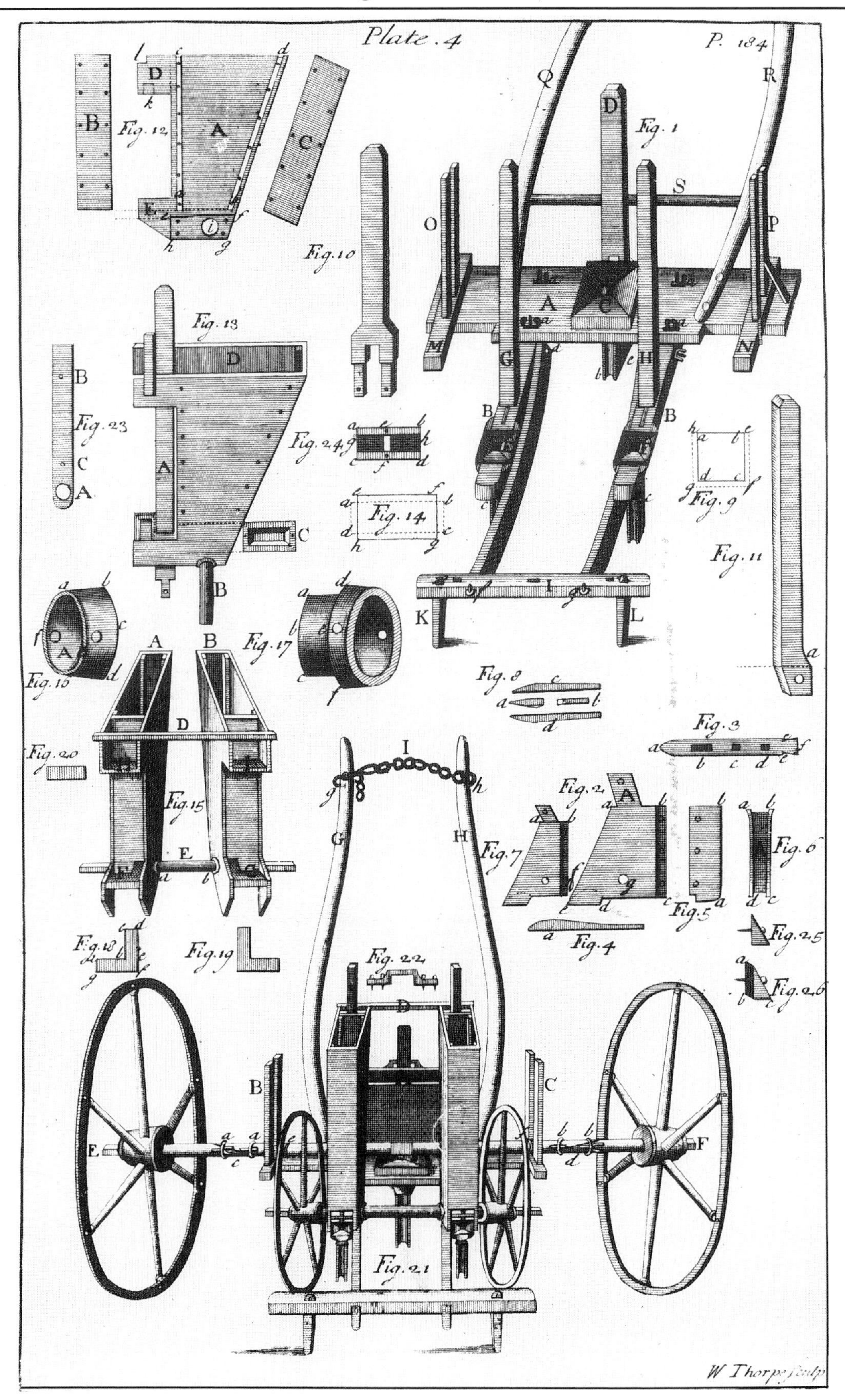
Plate. 4
P. 184
Fig. 1
Fig. 2
Fig. 3
Fig. 4
Fig. 5
Fig. 6
Fig. 7
Fig. 8
Fig. 9
Fig. 10
Fig. 11
Fig. 12
Fig. 13
Fig. 14
Fig. 15
Fig. 17
Fig. 18
Fig. 19
Fig. 20
Fig. 21
Fig. 22
Fig. 23
Fig. 24
Fig. 25
Fig. 26
W Thorpe sculp

Left **11** *The diagrammatic illustration of Jethro Tull's seed drill from the 1733 edition of* The Horse Hoeing Husbandry. *The drill was a small implement, built on a frame similar to that of a wheelbarrow. It could sow three furrows, the coulter for the middle one being set forward of the other two. A notched barrel (Fig. 3 in the illustration) was the means of guiding the seed from the seed box and into the dropper behind the coulter. Tull's inspiration for this mechanism had been the working of the organ he played in his parish church, and this was one of the major innovations of his drill for other inventors had used little more than a box with a hole in the bottom. [35/6775]*

12 *By the 1770s a number of small drills were available. Some were capable of sowing three or four furrows and were similar in pattern to Jethro Tull's original design. There were other single-furrow drills, such as this one offered by James Sharp of London, used mainly for sowing beans rather than corn. The small harrow attached at the back was a common arrangement. [35/7151]*

Left **13** *Northumberland was one of the regions where the seed drill was in use by the end of the eighteenth century. J. Bailey designed this seven-row drill for sowing all types of grain. The coulters were suspended independently so that they could ride over irregularities in the ground. [35/18543]*

THREE

The Rise of Agricultural Engineering: 1790–1850

When William Marshall, Arthur Young and others investigating the state of agriculture at the end of the eighteenth century considered the new implements of the farm they found a good deal to report. There were new types of ploughs, mole ploughs and cultivators being designed, all intended to give more thorough cultivation of the soil and to be of lighter draught than the old types of plough. Seed drills, after decades of neglect following Jethro Tull's work, were being redesigned and improved. There were the first threshing machines, too, usually small machines driven variously by hand, horse or water power. Not only were new tools and machines being devised, but some at least were being taken into general farming use. The demand for home-produced food and the high prices that resulted from the years of war with France between 1793 and 1815 encouraged farmers to turn to new implements to help increase arable production. But it was not by any means a universal trend. The seed drill, for example, was used little outside a few intensively cultivated districts in Suffolk, the Isle of Thanet and in Northumberland. The threshing machine similarly was taken up in parts of the northeast of England, the Scottish lowlands and East Anglia. However, these were clear indications that change was well under way.

By the time of the Great Exhibition of 1851 what the few were doing in 1790 was becoming commonplace. The section of the exhibition devoted to agriculture was one of the largest, a mark of the inventive ingenuity that had gone into the production of new tools for the farmer, from ploughs and seed drills to steam engines and mechanical bird scarers. The Great Exhibition was an extraordinary event, but there were regular showcases for the increasing number of new farm implements at agricultural shows, in particular those of the Bath and West of England Society and of the Royal Agricultural Society of England, whose first show had been at Oxford in 1839. It was from this time, the late 1830s, that demand for agricultural implements began to increase markedly. Agriculture was climbing out of the series of depressions with which it had been beset since the end of the Napoleonic Wars. An increasing population and the growing demand for food were making farming more prosperous and farmers in search of greater production turned to new implements and machines. The seed drill and threshing machine were now being used more widely throughout England and Wales. Intensive livestock husbandry required barns equipped with cake breakers, turnip cutters, chaff cutters and other machines which saved labour in preparing food for the animals.

The implements for which farmers of the 1840s were showing some enthusiasm were much improved upon those of Arthur Young's day. The report to the Board of Agriculture on the farming of Surrey, written in 1809, noted that one reason for threshing machines being little used was the farmers' view that 'from being ill-constructed or ill-managed [they] did their work ill'. Arthur Young held similar opinions about seed drills, and thought it remarkable that some survived crossing the field without shaking themselves to pieces. By the 1840s it was generally agreed that one of the major advances in the agricultural world since Arthur Young's day had been the improvement in the manufacture of implements. There were more implements on the market, and all were better made. Seed drills had become efficient tools, able to cope with variations in the texture of the soil and with uneven surfaces, and able to distribute seed with a fair degree of accuracy. Threshing machines were now more substantial, the largest able to thresh, dress and winnow about 350 bushels a day.

The relatively new technology of ironfounding made perhaps the greatest improvement to the quality of farm implements. Until the eighteenth century ploughshares and the tines of harrows had always been made of wrought iron beaten into shape by the blacksmith. Cast iron, one of the results of the revolution in ironmaking associated with the Darbys of Coalbrookdale early in the eighteenth century, was poured in liquid form into moulds of sand to make the ploughshares, feed mechanisms of seed drills, and other components of farm tools. The result was standardization: parts made to a regular shape and quality that could not be matched by the use of wrought iron. The price of iron was falling as the large blast furnaces in Shropshire and elsewhere were turning out greater quantities of pig iron, and this, for the maker of agricultural implements, meant that far more use could be made of iron. Hitherto, the great expense of iron had restricted its use to the cutting parts—the coulter and share—of the plough and the tines of the harrow. Almost all else was of wood, even the mouldboard. Cheaper iron extended its use to all the wearing parts of the new machines, the seed drills, threshing machines and cake breakers; and even to the more humble tools, the harrows and rollers.

Towards the end of the eighteenth century ironfoundries were being set up in agricultural districts. The cost of transporting pig iron and coal was coming down sufficiently for it to be worthwhile having the foundry close to the market for cast iron railings, gas lamp standards, pots and pans, and farm tools. Robert Ransome was one of those who did this, first at Norwich and then at Ipswich, a port more convenient for transport, where he established his foundry in 1789. Plough shares were among his staple lines of business, and although he was not the first to make shares of cast iron, he did introduce improvements in the quality of the material which have endured to the present day. The first was the tempering of the iron shares with salt water; the second was the 'chilling' of the shares, a process of hardening the under surface while leaving the top softer. The result of these improvements was that the shares were more malleable, less brittle and less likely to break against resistant ground; and they were self sharpening as the soft upper edge wore away more quickly.

Ransome took standardization a step further with a patent of 1808 for interchangeable plough parts. Instead of different types of plough being individually produced, Ransome was now able to offer standard plough bodies with a range of shares, coulters, wheels and other parts. It was a development that proved invaluable in meeting farmers' attachment to the traditional ploughs of their locality, for Ransome could produce the ploughs of almost every county in the land from a fairly small range of bodies and parts. Standard parts were appreciated by the farmers, for a lot of routine repairs and adjustments could now be done on the farm without having to send the implements to the blacksmith's shop.

On these foundations Ransome established the nineteenth century's leading agricultural engineering business. In 1851 his firm, which by now had passed to the second and third generations of the Ransome family, had 900 employees, more than twice as many as the nearest rivals. The Orwell works at Ipswich then covered nine acres, and was attracting eulogistic comment from visiting journalists. 'The range of manufacturing skill of Messrs Ransomes appears almost unlimited', wrote one whose report was widely syndicated in local and national journals. For their agricultural products included threshing machines, steam engines, ploughs, harrows, mills and cake crushers, all of which were 'manufactured by the Messrs Ransome in the best style of workmanship, and calculated to produce results which can only be attained by the highest and most experienced practical skill and science'.

Ransome's position as the leading agricultural engineering business of the time was the only justification for such comment for by 1850 there were many strong competitors, firms with good claims to fame and high standards for their products, and which were rapidly expanding. Several of these firms had been established within a few years of Ransome, many of them in East Anglia where some of the most advanced farming of the time was to be found. Garrett of Leiston in Suffolk was founded in 1778, and had grown to employ 300 by 1851; Burrell of Thetford, founded even earlier, in 1770, had 160 employees in 1851; and Hunt of Earls Colne, Essex, dating from 1825, employed 100. Elsewhere, two of the leading agricultural engineers established at this time were Hornsby of Grantham (1815) and Howard of Bedford (1811); each employed about 400 in 1851.

All of these firms were growing in the early part of the nineteenth century. Ransome by the 1830s had regular distribution of its products well beyond the confines of East Anglia into south Midland counties. What really made the agricultural engineering firms farm-household names were the improvements in communications: the railways, which carried the machines cheaply to distant buyers; and cheap postage and reduced costs of printing, which meant that illustrated catalogues could easily be distributed and links with sales agents maintained. Ease of distribution was a point emphasised in the firms' publicity. Howard's catalogues of later years often included maps to show that all railways led to Bedford. It is notable, too, that some of the firms which expanded most rapidly were founded as the railway network was being laid down. For example Clayton & Shuttleworth's Stamp End works at Lincoln, opened in 1842, was already employing 400 by 1851. Marshall of Gainsborough, established in 1848, had about 200 employees by 1861. And by this time the ploughs of Ransome and Howard, Hornsby drills, and Clayton & Shuttleworth threshing machines could be found throughout the country. These major firms were emulated to greater or lesser degree by numerous smaller iron works and engineering concerns and the manufacture of agricultural implements in the small workshop of blacksmith, ploughwright or wheelwright had now all but disappeared.

14 *The improved Rutland plough, one of the ploughs produced by Ransome of Ipswich. They introduced a plough of this name in the 1830s to a design by Richard Baker of Cottesmore. This is the remodelled version of 1843, replacing the wooden beam and handles with the elegantly curved iron pattern. The practice of making the land wheel smaller than the furrow wheel, not unknown in the eighteenth century, was now becoming common. [35/6815]*

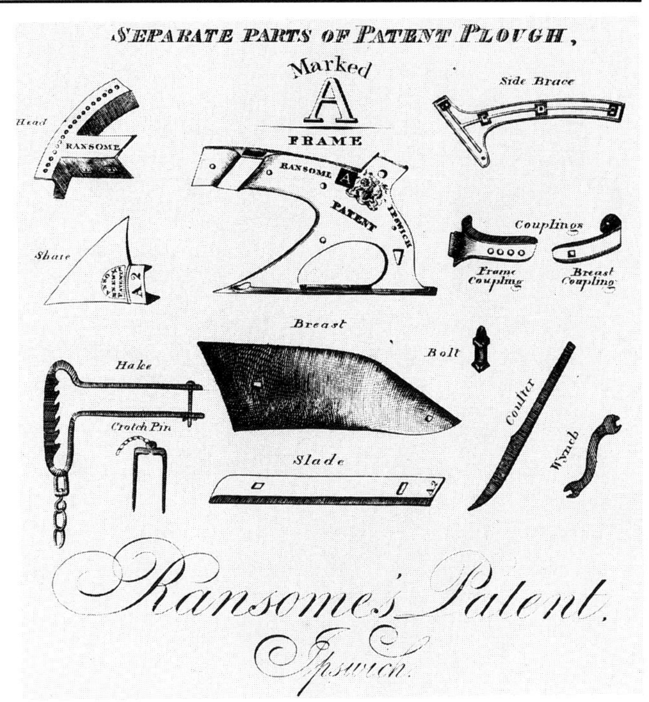

Right **15** *The catalogue of parts for one of Ransome's standard ploughs of the 1840s. The A model was a plough of the Suffolk type, having only one handle. [35/9041]*

Below **16** *Howard of Bedford became the principal rivals to Ransome as ploughmakers. Their 'Champion' ploughs of the later nineteenth century collected numerous awards at shows and ploughing matches. This was a double-furrow plough offered by the firm in their 1851 catalogue. It was of all-iron construction and distinctly more efficient in appearance than Mr. Duckett's plough of the 1770s. [35/26319]*

Bottom **17** *Early-nineteenth-century interest in improved drainage and deep ploughing resulted in a number of designs of subsoil plough. Garrett of Leiston, Suffolk, were producing this one in the 1840s. [35/5657]*

IMPORTANT AGRICULTURAL IMPLEMENT,

DISTINGUISHED FROM OTHERS BY THE NAME OF THE INVENTOR,

AND KNOWN IN THE COUNTY OF SUFFOLK, AS

BIDDELL'S SCARIFIER.

For the purpose of cultivating land under a variety of circumstances, and bringing it into a proper state of tilth, much more effectually and at less expence than can be done by the means generally employed for that purpose.

MADE OF DIFFERENT SIZES, BY

J. R. & A. RANSOME, IPSWICH.

18 *Tined cultivators, or scarifiers, were being produced to a number of designs at the beginning of the nineteenth century. In Devon there was a broad-shared version known as a tormentor which was reported in 1808 to be in quite common use. Arthur Biddell of Playford in Suffolk invented his scarifier, with two rows of tines that could be worked at varying depths, at about the same time. Ransome of Ipswich took up production of Biddell's scarifier in the 1830s, replacing the original wooden-framed version with all-iron construction and offering a variety of changeable chisel points and hoe shares. It was one of Ransome's most highly regarded products of the early 1840s. [35/12875]*

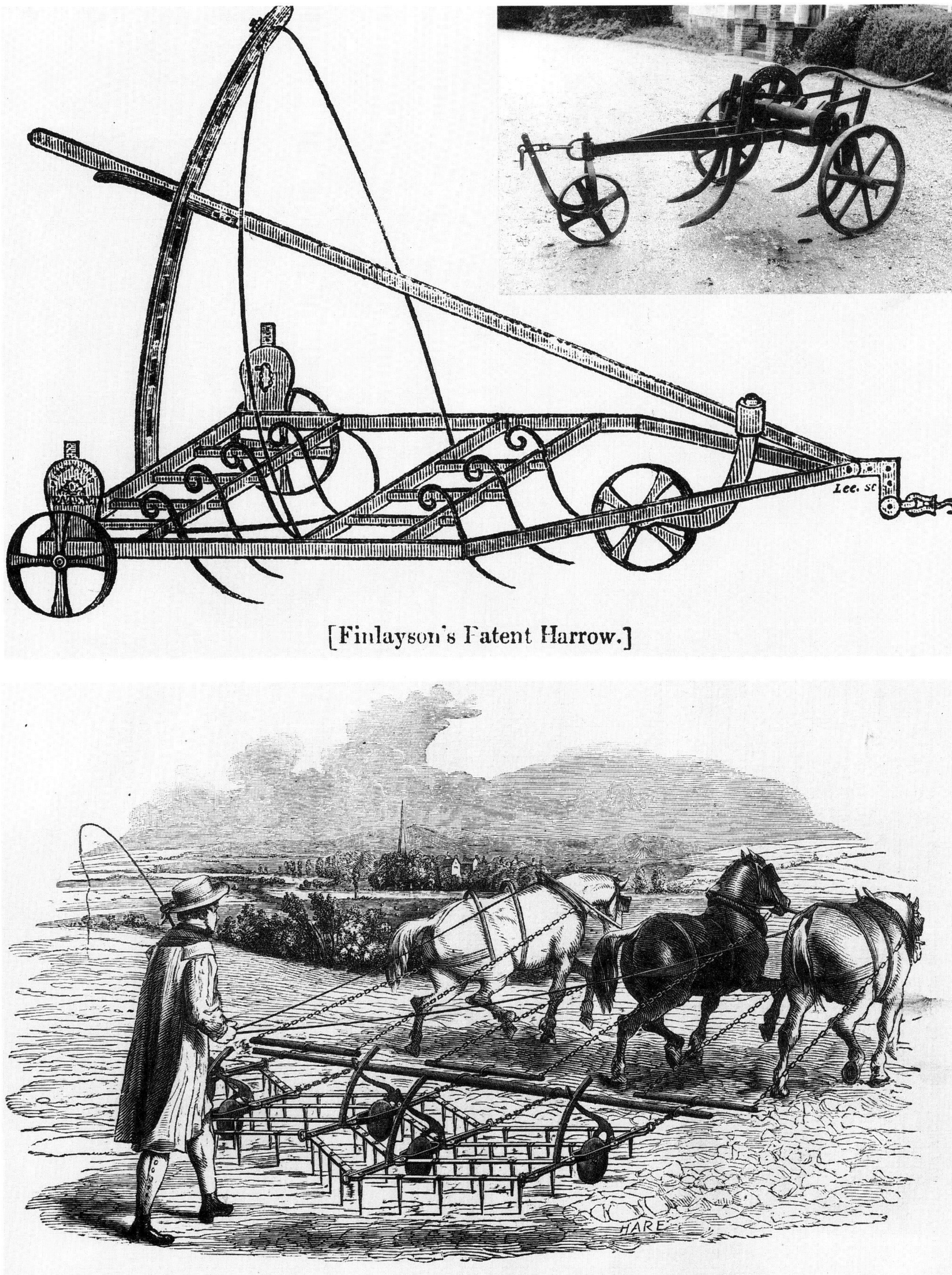

[Finlayson's Patent Harrow.]

COLEMAN'S PATENT EXPANDING LEVER HARROW.]

Top left **19** *Finlayson's patent self-cleaning harrow (bottom) was really a scarifier with finer tines than the type represented by Biddell's scarifier. His, and other implements similar to it, were popular implements for several decades. The inset photograph is a three-horse cultivator for mixed soils made in the late nineteenth century by Coleman & Morton of Chelmsford. [35/4275 & 60/291]*

Bottom left **20** *Coleman's expanding lever harrow, one of the variants on the rhomboidal iron link harrow produced in the 1840s. [5/137]*

Above **21** *One of the major benefits of cheaper iron was that relatively ordinary tools could be made to a reliable standard. Iron rollers replaced the variable shapes of tree trunks or stones used in earlier times, and new variations on the basic roller were devised. William Crosskill patented his clod crusher in 1841. As its name implies its main purpose was to break down the soil on ploughed land in spring cultivations. The roller was made up of a series of cast iron discs, serrated with both vertical and horizontal teeth. The discs, 30 inches in diameter, could spin independently of each other and chew up the clods. It was a most successful invention and widely used. [5/142]*

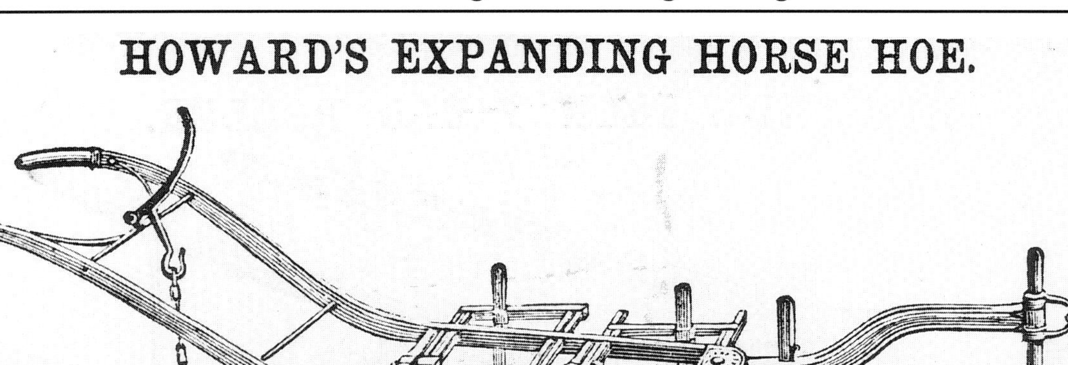

22 *Jethro Tull's seed drill had been but part of a larger scheme, horse hoeing husbandry, involving the regular stirring of the soil between crops planted in rows. Tull's horse hoe, however, was disregarded even more than his seed drill, and it was the 1840s–1850s before the horse hoe approached common use. Howard of Bedford produced the one shown top, one of the common types of horse hoe of the mid nineteenth century, advertised as suitable for one row of roots or three rows of corn crops. The one shown bottom was produced by Taskers of Andover in the second half of the nineteenth century, and was a similar type of horse hoe for weeding both roots and corn crops. The jointed construction of the body enabled it to be steered between the rows. The guide wheel at the front determined the depth to which the hoe would work and the harrow at the back dragged weeds to the surface. [35/18540 & 60/513]*

Top right **23** *Garrett's horse hoe of the 1860s, a larger implement than the Howard, suitable, according to the catalogue, for almost all crops. Up to 11 rows could be weeded with this hoe, covering up to ten acres in a day, at a cost, it was claimed, of 6d to 1s an acre. [35/4745]*

Bottom right **24** *The principles of Rev. James Cooke's seed drill, patented in 1782, had, wrote James Allen Ransome in 1843, 'been adopted in the construction of some of the most approved of the present day'. Particularly influential were the design of the seed box, the large carriage wheels, and the use of geared drive from the main axle to turn the feeding cylinder. [35/20325]*

Cooke's Drill

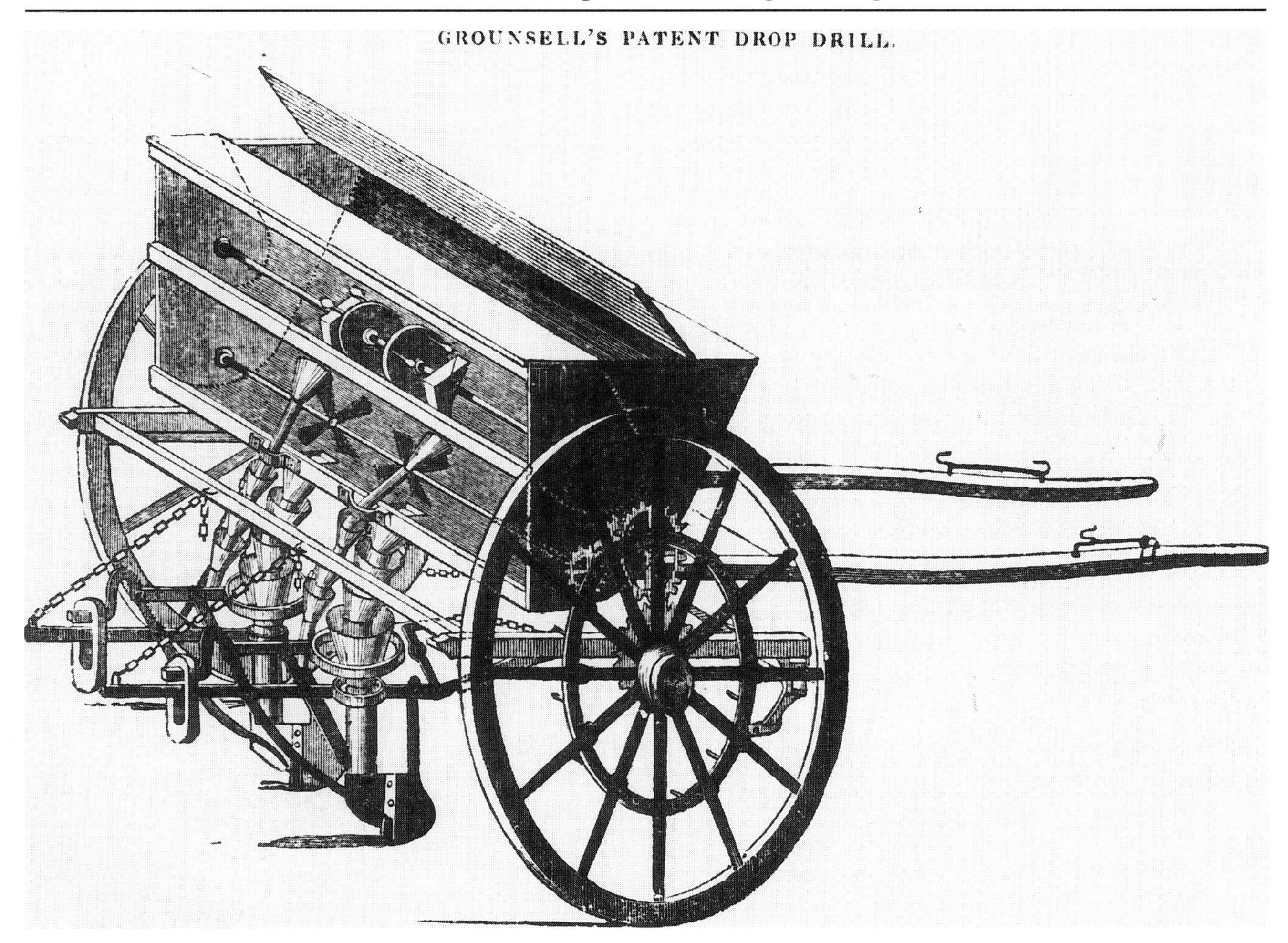

25 *Grounsell's patent drop drill of 1839 was one of the first seed drills able to sow seed at regular intervals along the row. The inner ring on the carriage wheel carried pegs which engaged the gears to open the valves on the seed delivery mechanism. The pegs could be set at different intervals around the ring to allow for the different spacing of different crops. Richard Hornsby & Sons introduced a drill with a drop feed, on a different principle, in the same year. [35/6737]*

26 *Garrett's general purpose drill, of which Philip Pusey remarked in his report on implements shown at the Great Exhibition, it was 'a very complete implement, capable of drilling, with or without manure, wheat, beans and turnips at the different intervals suited to those plants respectively, from 2 feet to 7 inches'. [35/5668]*

27 *The Suffolk drill was the single most important type of drill in British farming. It owed its origins to the work of James Smyth, of Peasenhall in Suffolk, who in the years immediately following 1800 made a number of improvements to the type of drill used in his district. These included the fitting of adjustable coulters, levers to lift them clear of the ground, and swing steerage. As subsequently developed, the principal features of the Suffolk drill came to be cup feed, lever lift mechanism and shoe coulters. This basic pattern, established by 1860, remained in use until recent times, while the company founded by James Smyth was the best known of all drill makers. The restored example shown below is a small, four row seed drill made by Smyths in 1928. [35/15236 & 6272]*

28 *The cup feed mechanism of a Smyth drill, from one of the firm's catalogues of the 1870s. [35/20470]*

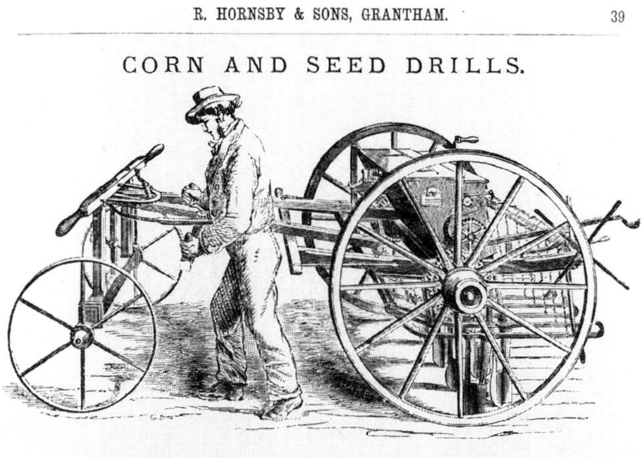

R. HORNSBY & SONS, GRANTHAM. 39

CORN AND SEED DRILLS.

THEIR SPECIAL POINTS OF MERIT ARE—

Loose Coulter Points, renewable like shares to Ploughs, and, being of chilled iron, very durable and cheap.

New screw Lifting Jack, to raise each end of Box.

All the Levers and Coulter Stalks of wrought iron, which are light, yet strong, and not subject to decay.

Continuous Seed Conductors of Patent Hose, very valuable for regular delivery, especially for light seed, beans, and pease.

All the Levers act independently, and have the same leverage in each, which can be regulated by the spare weights.

Strong and High Wood Carriage Wheels.

Moveable Shafts, and a full supply of Weights and Change Wheels.

R. Hornsby & Sons' Patent Steerage, with Rack and Pinion, giving the greatest command over the Drill, independently of horses, even when guided by a boy.

Left **29** *One of the common attachments to seed drills was a steerage mechanism to the front. It was a means of compensating for any lapses the horses might make from the prescribed straight line. The steersman walked along behind the forecarriage as shown in this illustration of one of Hornsby's drills made in the 1870s. [35/18551]*

Top right **30** *Drills incorporating shaped rollers for sowing turnips on ridges were being made early in the nineteenth century. By the middle decades of the century these drills were equipped with mechanisms to control the spacing and depth of sowing, and for adding manure, as shown in this example from the 1870s. The makers, James Coultas of Grantham, were established in 1863 and became one of the firms noted for seed drills. [35/20471]*

Bottom right **31** *By the 1840s it was usual for seed drills to be able to feed manure as well as to sow the seed. Separate manure spreaders were also available, either as drills or as general distributors for top dressings. Chandler's liquid manure distributors were among the machines featured at the Great Exhibition of 1851. [35/17594]*

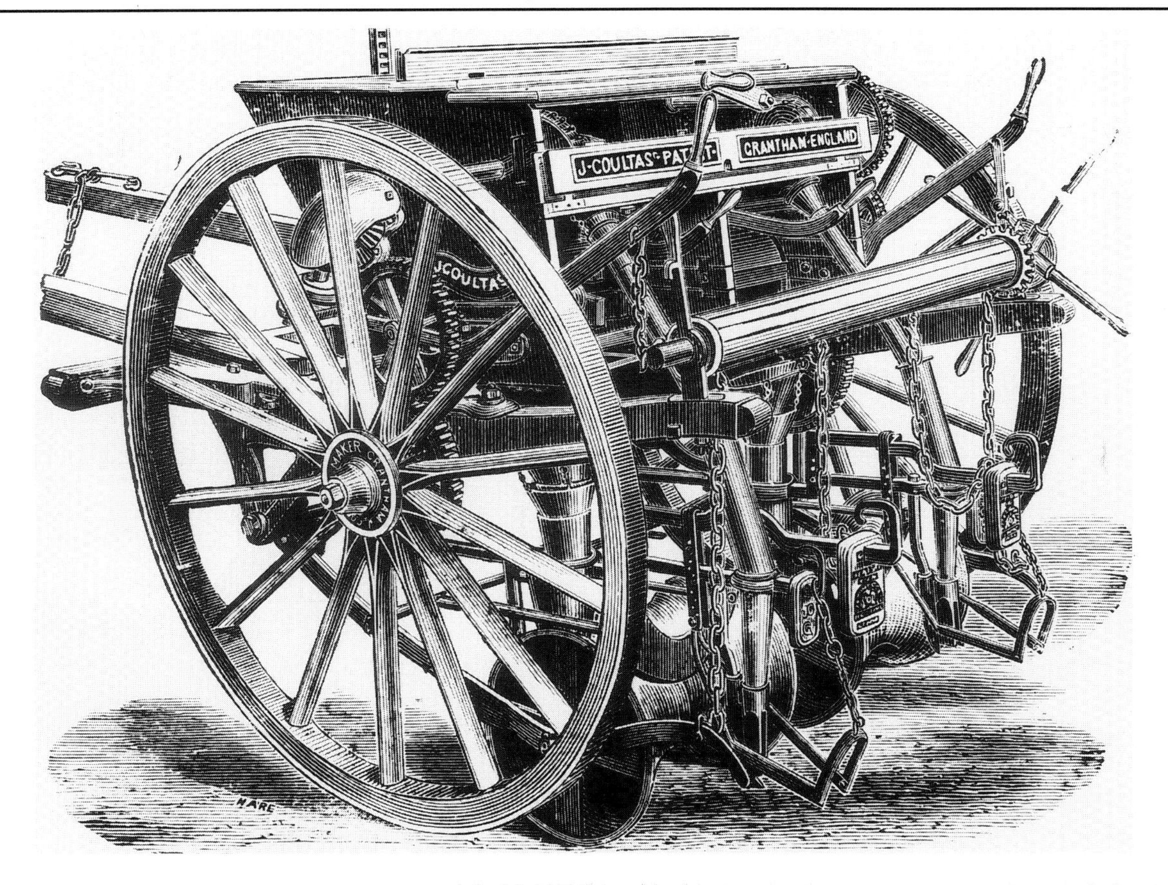

CHANDLER'S
PATENT
LIQUID MANURE DRILLS AND DISTRIBUTORS.

REPORTS OF THE JUDGES AT THE R. E. A. SOCIETY'S MEETINGS.

At YORK, 1848.

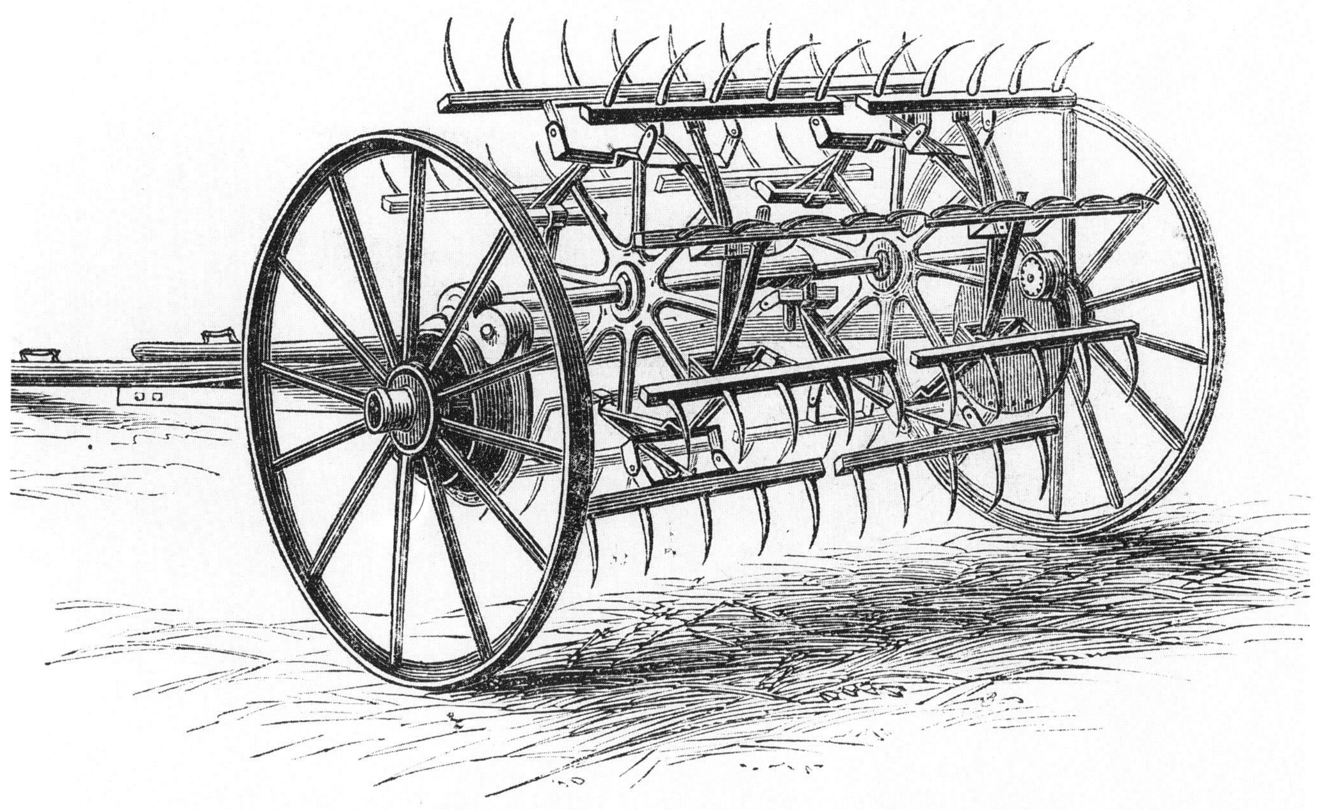

32 *The haymaking machine for turning the hay in the swath was invented some time in the early nineteenth century. An improved version was introduced about 1840 by Wedlake & Company, who had their iron works in Hornchurch and in the 'City of London Repository', where they sold a wide range of implements of their own and others' makes. Wedlake's machine had the tines mounted on sprung bars which would give on meeting a stone or similar obstruction and spring back into place once it was passed. The illustration is from Mary Wedlake & Company's catalogue of about 1849. [35/26274]*

33 *The first successful threshing machines were constructed in the late eighteenth century. Several people were involved, producing different machines, but the one who achieved lasting fame was Andrew Meikle. The principles of his design influenced the development of the threshing machine well into the nineteenth century. The corn was passed through a plain or fluted feeding roller to drums fitted with oak pegs which beat out the grain. Designs began to diverge from that configuration, but the Meikle principles remained popular in Scotland in the 1850s (Meikle was a native of the Haddington area). This illustration of a threshing machine 'on the Scotch principle' is from an agricultural encyclopaedia of 1855. Like most of the first threshing machines it is shown built permanently into a barn.* [35/26277]

RANSOME'S HAND-THRASHING-MACHINE.

Left **34** *While most early threshing machines were permanent fixtures in barns, small, hand-powered machines began to appear in the early nineteenth century, such as this one illustrated in the* General View of the Agriculture of Berkshire *by W. Mavor. Despite the demure expressions on the girls' faces the machines required considerable energy to operate. Mavor described this one as 'more curious than useful', and considered that there was little saving over threshing by flail. [35/8469]*

Bottom left **35** *James Allen Ransome declared that 'manual force is not so inapplicable to this object as most have represented it to be'. His company's threshing machine, to be worked by four men, was commended by judges at the Royal Show at Liverpool in 1841. Ransome's views were amply borne out by the fact that hand-powered threshing machines continued to be made into the present century. [60/9256]*

36 *Garrett's improved portable threshing machine of the 1840s, with which the Suffolk firm won prizes at many agricultural shows. It operated with a main cylindrical wire drum and a second set of pegs to catch ears of grain that had slipped through the drum. [5/732B]*

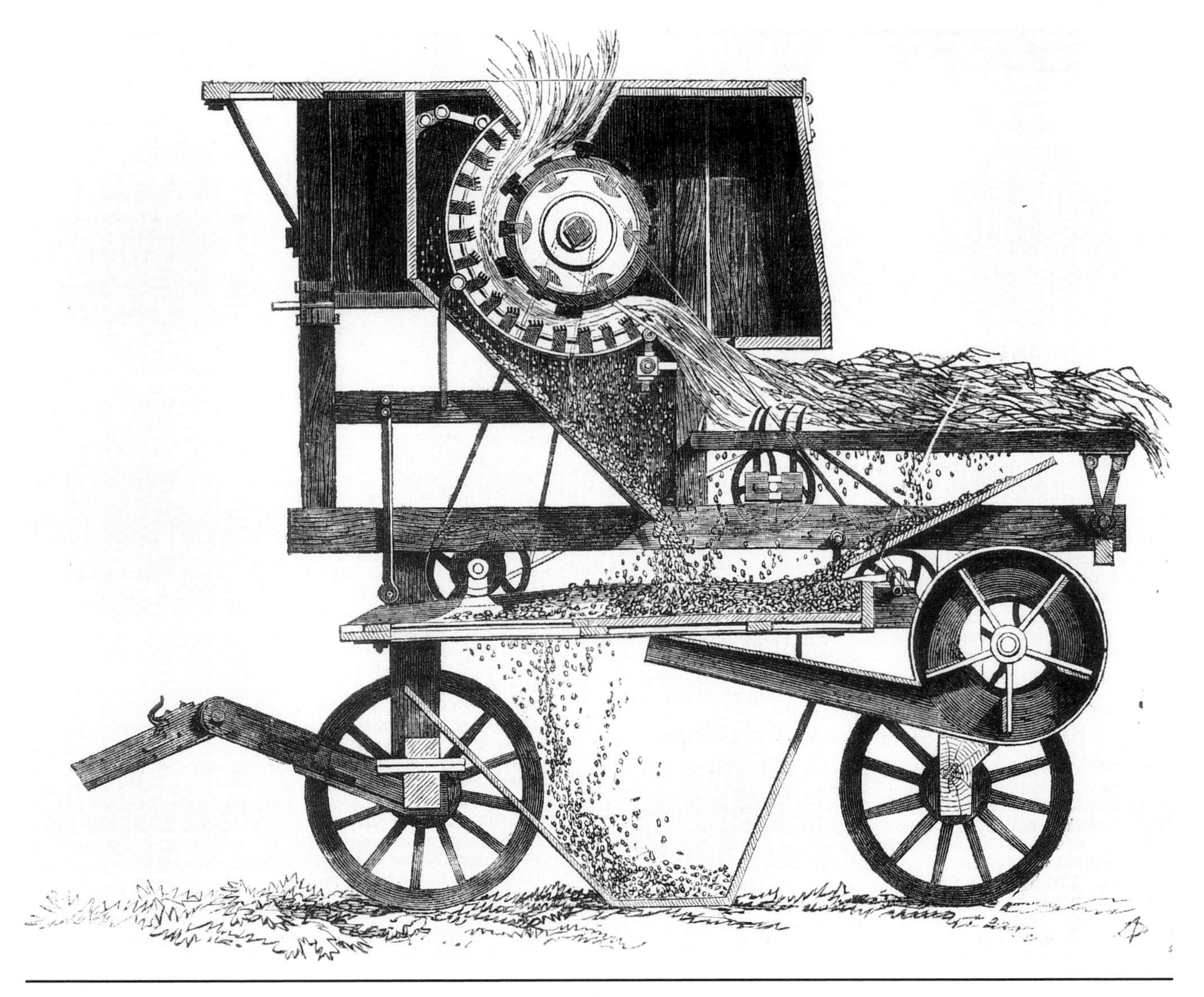

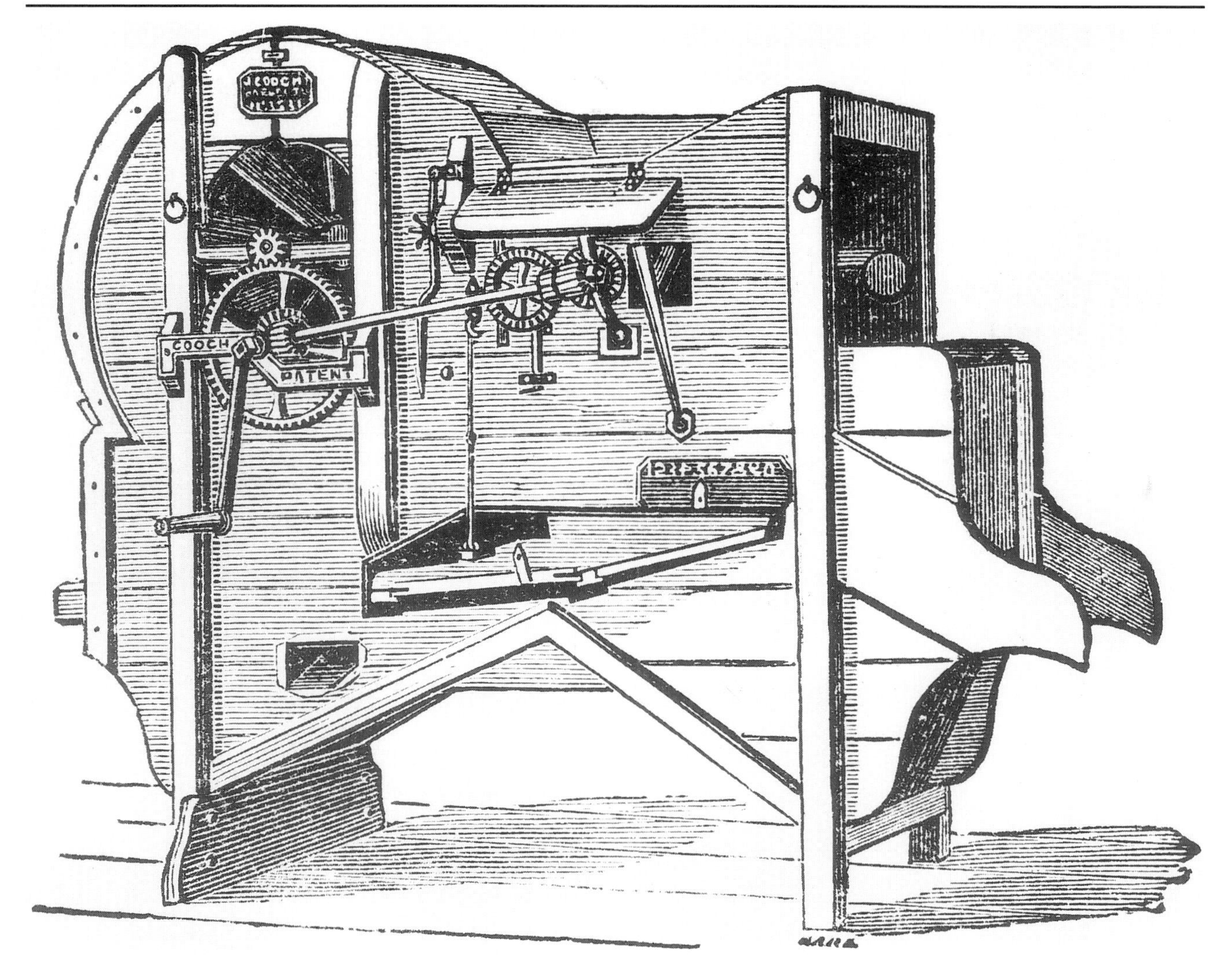

37 *Cooch's winnowing machine. A basic tool for winnowing had been introduced from Holland in the eighteenth century. This was a number of canvas sails attached to a hand-operated cranking mechanism by which they were made to revolve and create the wind to blow away the chaff. From the beginning of the nineteenth century the enclosed box type of winnower appeared. Inside the box was a drum with toothed wheels which acted as a fan on being made to revolve quickly. The corn, fed through a hopper at the top, was made to pass through a series of sieves which cleaned it of all the other dust and stones. J. Cooch of Northampton took out one of the earliest patents for this type of machine, and the more sophisticated version of the 1840s, shown here, was a firm favourite.* [35/17968]

Left **38** *Clayton & Shuttleworth's combined threshing, riddling and winnowing machine, from the catalogue for the Great Exhibition, 1851. The successful combination of threshing, winnowing and dressing into one machine was achieved in the late 1840s, Charles Burrell & Sons producing the first in 1848. Although Clayton & Shuttleworth here advertise a three-horse version of their threshing machine, the new larger machines were really dependent on the extra power of the steam engine. [35/11090]*

39 *A two-horse gear manufactured by Barrett, Exall & Andrewes of Reading, from the Great Exhibition catalogue. Horse power for driving threshing and other machines was extremely important in farming until well into the twentieth century, and the horse gear an essential part of the farm's equipment. The type of horse engine shown here was a product of the new industry of iron founding at the beginning of the nineteenth century. [35/23414]*

40 *Horse gears continued to be made in the twentieth century, notably by Hunts of Earls Colne, Essex, whose two-horse 'Express' gear with three-speed intermediate gearing was one type produced in the years before the First World War. It is shown at top coupled to a grinding mill. The restored example shown below is a two-horse gear with intermediate motion, made about 1910. The horses walked about $2\frac{1}{2}$ circuits per minute, and the gearing transferred this motion to drive the belt shaft at 400 revolutions per minute. It is here coupled to a combined crushing and grinding mill made by Bentalls. [35/26402 & 60/13935]*

CHAFF-CUTTING ENGINES.

J. R. & A. RANSOME,

PATENTEES FOR IMPROVEMENTS ON PLOUGHS,

IPSWICH,

41 *The more intensive livestock husbandry of the times was reflected in the number of machines that became available in the 1830s and 1840s for preparing cattle food. Chaff cutters had been invented in the 1770s. Ransome's chaff cutter of the 1840s, illustrated here, was one of the prize winners of that period. [35/6816]*

Garrett's Rape and Linseed Cake-Crusher.

Right **42** *Garrett's rape and linseed cake crusher, a two-hand machine of the 1840s. Oilseed cake was supplied in large blocks which had to be broken down. The guillotine knives used at first were superseded by the cake crushing machines. [35/4313]*

Right **43** *In the eighteenth century turnips for feeding to cattle were sliced by ordinary knives. Turnip cutting machines were devised about 1820, and by the 1840s several variations were available. Gardener's root cutter was generally regarded as one of the best at least into the 1860s, the date of this illustration. [35/6883]*

Gardener's Turnip Cutter with Iron Frame.

44 *The rise of the agricultural engineering industry: Ransomes & Sims' stand at the International Exhibition, London, in 1862. [35/18402]*

FOUR

The Steam Age in Farming: 1850–1940

The middle decades of the nineteenth century have often been characterized as a period of Victorian optimism, when Britain 'led the world', especially in the realms of manufacturing and commerce. There seemed no limit to the advances that could be achieved. It was a spirit that influenced agriculture no less than other spheres of activity. With both demand for food and prices for farm produce generally moving upwards, farming, it was argued, had everything to gain from adopting all the advances that science and technology had to offer. Indeed, agriculture should as far as possible emulate manufacturing industry and pursue greater mechanization in order to 'make the business of the farm approximate more closely to that of the factory'.

Central to this line of argument was the replacement of animal power by mechanical power. The major industries were using steam engines to an increasing extent; so should agriculture. A great deal of ingenuity was directed towards the design of steam engines for agriculture, especially for steam-powered cultivating and ploughing. Most of these last were failures, but steam did win its place on the farm and several of the leading agricultural engineering companies, including Clayton & Shuttleworth and Charles Burrell & Sons, made their names as steam engineers.

Steam power did not come to the farm much before the 1830s, although there had been a few engines in use driving threshing machines by 1800. There was little incentive, in the economic conditions affecting agriculture before the 1830s, to invest in a steam engine. In addition, high pressure steam engines of modest dimensions suitable for installation in a barn were not readily available until then, the engines used in farming previously having been relatively inefficient. The first steam engines in agriculture were stationary, built into the barn. They were used mainly on large farms and in estate yards where they drove such things as sawbenches as well as threshing machines.

Stationary engines retained a place in agriculture until the end of the steam era and many of the agricultural engineering companies, great and small, made them. Their use on the farm, however, was limited: most of the agricultural engineering companies' output was sold overseas or to non-agricultural use. Instead, for farm work engines on wheels were preferred: engines that could be taken to work in the barn, in the stackyard or out in the fields. The fact that machine threshing, one of the major tasks suitable for steam power, was often undertaken by contractors travelling from farm to farm served to make this consideration even more important. So it was that the characteristic agricultural steam engine was to be the portable engine. It had a locomotive type tubular boiler and firebox, a single or compound cylinder, and a very tall chimney—6–8 feet high—to carry the steam and smoke well away from the stackyard. Most of the portable engines used in this country were of medium power—6, 8 or 10 nominal horsepower—but engines of up to 20 nhp were in regular production, and by the twentieth century some compound-cylindered engines of 30 nhp were being made, though mainly for export.

A number of engineers developed the portable engine simultaneously. Ransome of Ipswich caused a stir at the Royal Agricultural Society's show in 1841 when they brought the first portable steam engine to be exhibited at these shows, an engine produced in conjunction with the Birmingham Disc Steam Company. Before that, in 1839, William Howden, of Boston (Lincs), had produced a portable engine which he demonstrated locally at the Lincolnshire Agricultural Show, while a fellow Boston engineer, William Tuxford built his first portable in 1842. Once successfully demonstrated, the portable engine caught on quickly. It was noted that a feature of the Royal show of 1847 was the increased number of steam engines, with seven firms exhibiting. Sales were brisk as well: there were said to be as many as 8,000 portables in use in 1851. This was probably an overestimate, but Clayton & Shuttleworth, founded in 1842, sold over 200 portable engines in 1851, and by 1857 had made a total of about 2,400. During the next 70 years or so, Clayton, Marshall, Ransome, Ruston, and all the other firms made vast numbers of this type of engine for markets at home and overseas.

The portable engine could be moved, but could not move itself, requiring instead to be pulled by horses from farm to farm. Ransome again claim the credit for the first self-moving agricultural steam engine, for they demonstrated a self-propelled version of their portable engine at the Royal show of 1842. In 1849 they exhibited the 'Farmer's Engine', a more promising engine built for them by E. B. Wilson's Railway Foundry, Leeds, but it took a few more years before the problems of building engines that could cope with the rough roads on and off the farm were overcome. Charles Burrell achieved this first, in 1856, with Clayton & Shuttleworth, Tuxford, and Ransome following closely behind.

Once established as a successful engineering proposition the traction engine gained its place in agriculture, being especially favoured by the contractors who could dispense with horses for hauling equipment between farms. They were generally called by the manufacturers 'general purpose agricultural engines', but in practice the purpose to which they were usually put on the farm was driving the threshing machines, and associated equipment, such as straw balers, and clover hullers. They could be used for haulage, but for most purposes it was as convenient for the farmer to employ a horse and wagon. Traction engines, for the most part, were made in the same range of power as portable engines, between 6 and 12 nominal horsepower, that being quite adequate even for the larger threshing machines made in the late nineteenth century.

The traction engine was not used to pull ploughs and other cultivating implements across the fields in the way that a tractor does, for it was too heavy. The lightest of standard agricultural engines weighed about 6 tons, and most were of 9 or 10 tons, weights which were likely to pan the soil. It was this problem which dogged all efforts to introduce steam power for ploughing and cultivating, although many attempts were made. In the end the solution adopted was that of keeping engines at the side of the field hauling implements across by a cable. John Fowler brought the most straightforward system to fruition during the late 1850s. He used a pair of engines (or one engine and a moveable windlass), one each on opposing headlands, and they passed the implements between them across the field. The company founded by Fowler dominated the market for steam cultivating equipment until it was ousted by tractors in the twentieth century.

Steam cultivation was impressive. The engines were immensely powerful, producing four or five times the 12 or 14 nominal horsepower at which they were officially rated. With that amount of power substantial implements could be used: five or six furrows were usual for the ploughs, some larger still. This great power proved invaluable for deep cultivation, for breaking up fallows, especially in dry seasons, and for cultivating heavy clay soils. The high capital cost—about £1,600 for a set of Fowler ploughing equipment in the 1880s—ensured that steam ploughing was undertaken almost exclusively by contractors. The fact that British fields were so often small, of irregular shape, and otherwise awkward for manoevring steam engines was a further reason for the continued dominance of the horse and single furrow ploughs on most farms.

The second major development in farming technology of the late nineteenth century was the mechanization of harvesting, the most important part of arable farming that in 1850 was still being carried out entirely with hand tools. The Great Exhibition of 1851 marked a turning point, for two American exhibitors, Cyrus McCormick and Hussey, had on display reaping machines which aroused a considerable amount of interest. They were by no means the first reaping machines to be seen in this country—there had been some made earlier in the century, most notably Patrick Bell's reaper invented in the 1820s—

but these American ones were the first demonstrated to be reliable and efficient in operation. They appeared on the scene at just the right time, for farmers were beginning to experience difficulties with the supply of labour for harvesting, and so were becoming amenable to labour-saving machines. Slowly at first, but after 1860 with gathering pace, mechanical reaping gained popularity. By 1871 about 25 per cent of the harvest was cut by machine, and about 80 per cent by 1900. The process of mechanization was taken a stage further during the 1890s when farmers began to take up the self-binding reaper, which tied the cut corn into sheaves where the ordinary reaper left it in a swathe across the field.

The hay harvest, too, became more mechanized. Mowing machines, similar to the corn reapers but with blades suitable for cutting grass, came onto the market in the 1850s. The Royal Agricultural Society's first major trial of hay mowers was in 1857, when Clayton & Shuttleworth's American Eagle mower was awarded the first prize. Cutting the hay by machine became an established part of farming during the next few decades, although mowing by hand was by no means eliminated by 1914. Haymakers, hay tedders and swath turners all became more popular, and a large number of manufacturers had these machines on offer. Elevators for making the haystack were another introduction of this period.

The new reapers and binders had a significance beyond the introduction of machines into harvesting, for they were the first major agricultural implements to be introduced from abroad. However much the lineage of McCormick's reaper could be traced back to Patrick Bell's machine, the fact was that they were American products on display at the Great Exhibition. They became naturalized, adapted to British conditions, as it was British firms who took up the manufacture of the Hussey and McCormick machines under licence, and those firms introduced their own modifications and improvements. Thus it was that Samuelson of Banbury, Hornsby and the 'Albion' became names better-known for harvesting machinery than McCormick before the First World War. But it was not long before North American companies were looking for a more direct involvement in the British market, and the main competition from imports before 1914 came in the market for ploughs. Imports of American steel chilled ploughs began in the 1870s, and by the 1890s they had established a firm foothold in a British market for agricultural implements that was generally rather inactive. As an indicator of future trends this development was important.

45 *The farm beginning to approximate more closely to a factory. An illustration from Samuel Copland,* Agriculture Ancient and Modern *(1866), of a barn filled with machinery, including a threshing machine, chaff cutter, mill, root cutters and cake breakers; and all driven by steam, a portable engine in the top right corner. [35/8212]*

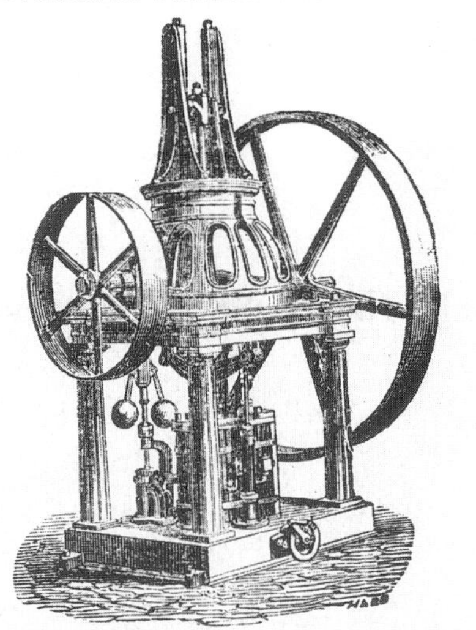

46 *Tuxford's steeple, vertical steam engine, from the catalogue for 1872. It was constructed on a cast iron sole plate to be mounted on a plinth of stone or concrete. The twin flywheels allowed for flexibility in the operating speeds of the barn machinery, for which this was advertised as suitable motive power. The price for the four-horsepower engine was £150. [35/26381]*

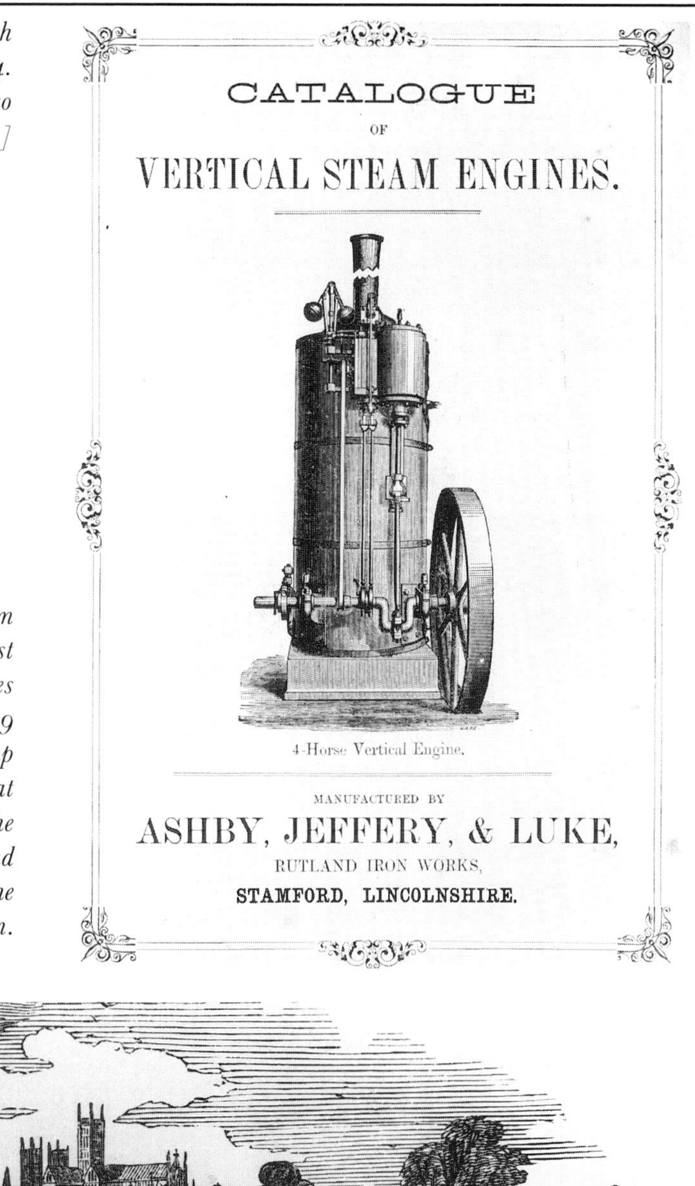

CATALOGUE

OF

VERTICAL STEAM ENGINES.

4-Horse Vertical Engine.

MANUFACTURED BY

ASHBY, JEFFERY, & LUKE,

RUTLAND IRON WORKS,

STAMFORD, LINCOLNSHIRE.

Right **47** *A more common type of vertical steam engine with multi-tubular boiler made by Ashby, Jeffery & Luke, 1874. The four-horsepower version of this cost £110. This firm also offered a portable version of their vertical engines. [35/26380]*

48 *Clayton & Shuttleworth single cylinder portable steam engine of 1851. This company was one of the largest manufacturers of agricultural steam engines. Their portables in 1851 were available in power sizes ranging from 3 to 9 nominal horsepower, at prices from £135–£255. The 5 hp engine was reckoned capable of threshing 35 quarters of wheat a day. Jabez Hare, the artist and engraver of many of the illustrations of agricultural equipment in catalogues and magazines, has shown Clayton's engine and threshing machine against a backdrop of the city and cathedral of Lincoln. [35/26321]*

Left **49** *Marshall & Sons, of Gainsborough, founded in 1848, was another of the major manufacturers of steam engines for agriculture. One of the company's single cylinder portable engines, available in power ratings from $1\frac{1}{2}$ hp to 12 hp; late 1870s. [35/13069]*

50 *Traction engine by Charles Burrell & Sons, Thetford, 1881. Single cylinder engines of 6 hp–12 hp intended for use as agricultural or road locomotives. [35/17120]*

MR. HEATHCOAT'S STEAM-PLOUGH.

51 *This unlikely-looking machine was the first practical steam plough. It was the work of John Heathcoat, a lace manufacturer, and Josiah Parkes, a drainage engineer. It worked on the principle of indirect traction: the plough was hauled across the field by cable on a winding drum powered by the engine on the headland. This engine had endless tracks of considerable size to spread its weight, and to demonstrate their effectiveness the two partners demonstrated their machine on the boggy land of Red Moss, near Bolton le Moors in Lancashire in 1835. The trials were reasonably successful, but the project was a failure financially and was abandoned. [35/7543]*

352 THE ILLUSTRATED LONDON NEWS [Oct. 3, 1857

STEAM CULTIVATION: CROSSKILL'S ROMAINE CULTIVATOR.

52 *Diggers or rotary cultivators were canvassed as the most appropriate form that steam cultivation should take, and several inventors produced machines on those principles. Robert Romaine's steam cultivator was the most successful of those in the mid nineteenth century, produced as a self-moving steam-powered machine by Crosskill of Beverley in 1857. However, doubts about the weight of the engine on the soil, and farmers' scepticism about alternatives to the plough, meant that development went no further. [35/20882]*

HOWARDS' NEW PATENT STEAM PLOUGHING APPARATUS, AS AT WORK.

53 *The principal alternative to the Fowler system of steam ploughing was this, known as the roundabout system. An engine in the corner of the field provided power for the winding drum, the cable being guided round the field by pulley wheels at the corners. William Smith, a farmer of Woolstone, Buckinghamshire, was the most prominent of those who, during the 1850s, made this system a practical proposition; Howard of Bedford were the largest manufacturers of such equipment. The great advantage of the roundabout system was that it required only one engine, and an ordinary portable engine at that, rather than a specially equipped ploughing engine. But its drawback was the tiresome and back-breaking task of moving the heavy guide pulleys as the plough completed each furrow. So, by the 1880s the roundabout system had lost the contest with Fowler's. [35/7548]*

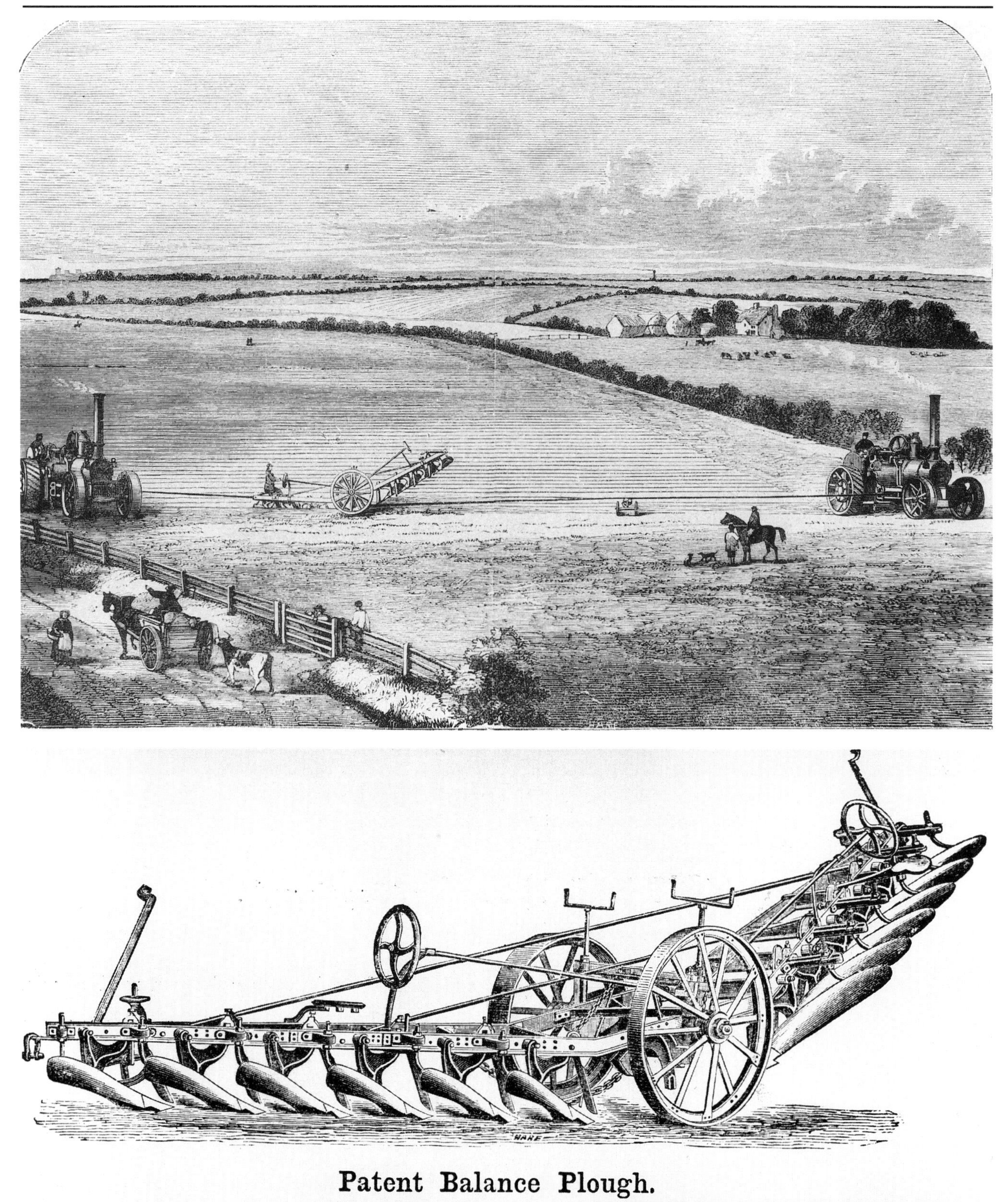

Patent Balance Plough.

Top left **54** *Fowler's double-engine system of steam ploughing, a scene engraved by Jabez Hare for the firm's catalogues of the 1880s.* [35/26279]

Bottom left **55** *The Fowler balance plough, a six-furrow version from the catalogue for 1874. At the end of each passage across the field the plough was turned about the axis of the central wheels to reverse it. The double set of shares meant that the plough worked on the turnwrest principle, turning all the furrows in the same direction.* [35/18544]

56 *As well as the antibalance ploughs Fowler produced a range of cultivating implements for use with steam cable engines, including scarifiers, harrows, rollers and seed drills, and all much more substantial than any implement available for horse traction. This is a two-shaft reversible discer at work in the Lincolnshire fens in 1926. The engine is one of Fowler's BB class compound ploughing engines.* [35/10983]

Top left **57** *The class BB1, introduced in 1917, represents the final generations of Fowler's ploughing engines built for the home market. They had compound cylinders, and were rated at 16 nominal horsepower. Orders by the government during the First World War amounted to 159 of these engines. No. 15142, seen outside Fowler's works in Leeds, was delivered in 1918.* [35/23726]

Bottom left **58** *The most celebrated of the late nineteenth-century types of steam cultivating equipment was the Darby broadside digger. Thomas Darby of Pleshey, Essex, made his first digger in 1877 and produced a number of versions during the 1880s and early 1890s. All worked on the same principle. A series of forks arranged along one side of the double-boilered steam engine was driven by shafts and gears linked to the crankshaft of the cylinder. On the other side of the engine was an outrigger to give the machine stability. The pair of wheels under each boiler gave further support, and could be turned through 90° for travel along the road. The Darby digger won medals at Royal Shows and praise from farmers who used it—but very few orders.* [35/12006]

59 *Early in the twentieth century, to meet the challenge that tractors were more versatile, steam engine builders produced some light steam tractors intended for direct haulage of cultivating implements. Wallis & Steevens of Basingstoke produced their 4½ ton compound tractor, photographed demonstrating its abilities with ploughs fore and aft.* [35/26280]

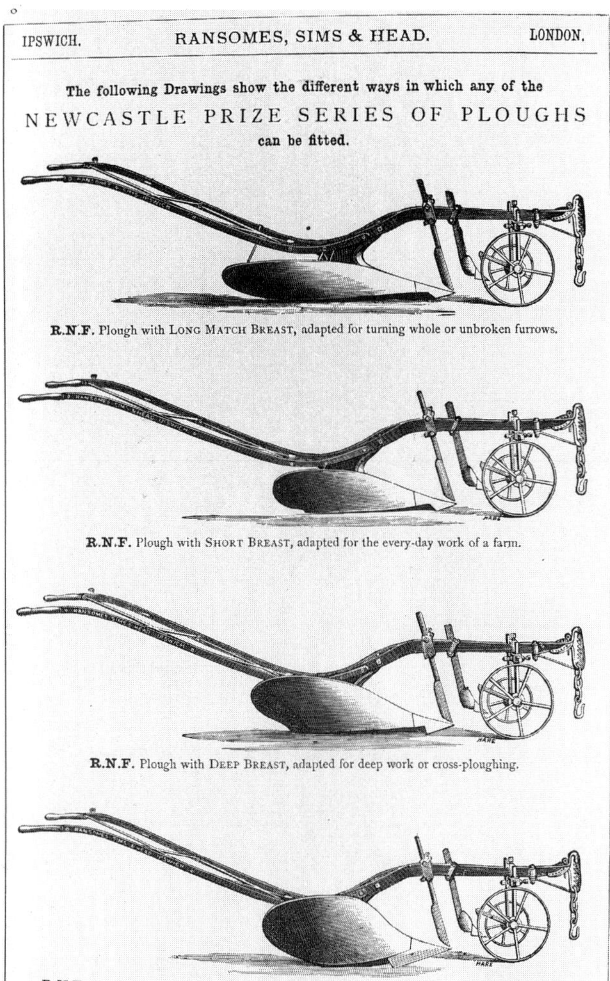

IPSWICH. RANSOMES, SIMS & HEAD. LONDON.

The following Drawings show the different ways in which any of the

NEWCASTLE PRIZE SERIES OF PLOUGHS

can be fitted.

R.N.F. Plough with LONG MATCH BREAST, adapted for turning whole or unbroken furrows.

R.N.F. Plough with SHORT BREAST, adapted for the every-day work of a farm.

R.N.F. Plough with DEEP BREAST, adapted for deep work or cross-ploughing.

R.N.F. Plough with WIDE BREAST, adapted for turning a furrow 14 inches wide, 6 inches deep.

Left **60** *Ransome's new RNF plough carried off the top prizes at the Royal Agricultural Society's show at Newcastle in 1864, and for that reason was thereafter known as the Newcastle plough. It was assiduously promoted by its makers through entries at shows and ploughing matches, which helped to maintain its position as one of the major types of horse-drawn plough. It continued to be made until the 1940s. These were four of the many versions of the Newcastle plough advertised in Ransomes, Sims & Head's catalogue for 1869. [35/8120]*

61 *The British agency for Oliver steel chill ploughs was set up in 1886, and they quickly became the best sellers of the American ploughs. The company's advertising modestly declared that Oliver ploughs 'in a few years have completely revolutionised the plow trade of Great Britain and Ireland'. The success of the American ploughs brought forth imitations from British companies, among them Ransome and John Cooke of Lincoln, whose ploughs were claimed to be more suited to British conditions. All, however, followed the basic design of the American plough, with the chunky wooden beam and single wheel. (A contemporary illustration from the company's leaflet can be seen on the back jacket, top.) [60/13335]*

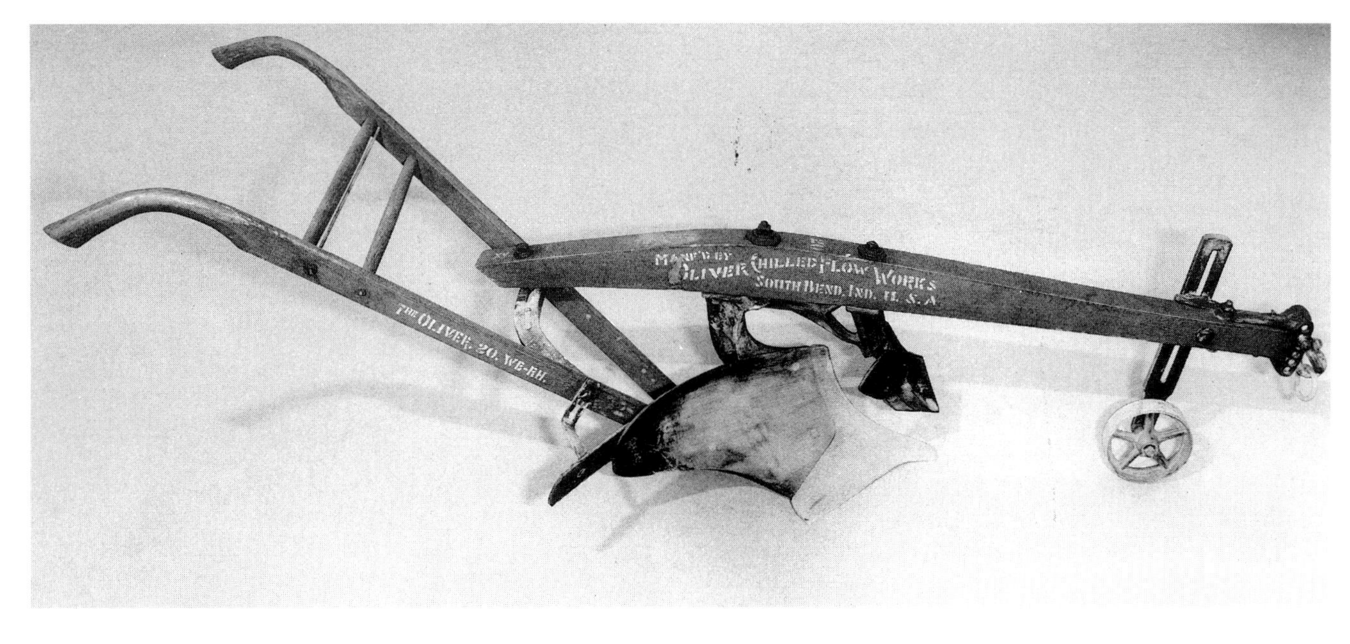

HUXTABLE'S PATENT "PERFECTION" ONE-WAY PLOUGH.

THE design and construction of this plough combine strength and durability with lightness and adaptability. The beam is made of solid drawn steel tube, practically unbendable. The lever necks carrying shares are adjustable to suit all conditions of land. The method of operating front wheels for turning at end of furrow, is simple and reliable, never failing to act at the right moment. The draught of plough is reduced to minimum by use of separate anti-friction chilled heelpiece.

Can be made with 1 Set of Ordinary Plates, or as a Digger only, or with 2 Sets of Plates, as shown.

62 *During the nineteenth century several manufacturers sought to bring up to date the principles of the turnwrest plough in turning all furrows the same way. The design of the turnover plough, in which two sets of plough bodies were revolved in the horizontal plane around the beam, was settled by a patent taken out by John Huxtable of Barnstaple in 1889. The version offered by Huxtable in 1920 is illustrated here. [35/15240]*

63 *A two-furrow balance plough at work in the 1930s. Less popular than the turnover plough was the one way balance plough, which was really a smaller version for horse power of the reversible plough used in steam ploughing.* [60/8799]

This implement is intended to follow the Plough, and by its use a firm bottom for the seed is obtained, the ravages of the wire-worm are checked, and on many lands better crops of wheat and oats are obtained than by other modes.

				£	s.	d.
Two-wheel Round Spindle Presser,	32 inches diameter	...		5	15	0
Three ,, ,, ,,	32	,,	...	8	0	0
Two ,, ,, ,,	36	,,	...	7	10	0
Three ,, ,, ,,	36	,,	...	9	10	0
Two ,, ,, ,,	42	,,	...	8	10	0
Three ,, ,, ,,	42	,,	...	11	10	0

The above Pressers are fitted with hard boxes cast in the centres of the wheels, which prevent their wearing.

				£	s.	d.
Two-wheel Square Spindle Presser,	32 inches diameter	...		7	7	0
Three ,, ,, ,,	32	,,	...	10	0	0
Two ,, ,, ,,	36	,,	...	8	8	0
Three ,, ,, ,,	36	,,	...	11	0	0
Two ,, ,, ,,	42	,,	...	10	0	0
Three ,, ,, ,,	42	,,	...	12	12	0

The wheels of these Pressers are fitted on a square spindle, which revolves in brass bearings.

If desired, a small drill box can be attached to either size for depositing seed behind the presser.

		£	s.	d.
Extra fitted to a Two-wheel Presser	... ...	£5	0	0
,, ,, Three ,, ,,	... ...	6	0	0

WATERLOO IRON WORKS, ANDOVER, HANTS.

64 *The seam press, or land press, was developed during the first half of the nineteenth century, and gained popularity during the century's middle decades. It followed the plough, used to roll the base of the open furrow, and so produce a firmer seed bed. Several manufacturers had these implements on offer, among them Tasker & Sons of Andover. [35/26377]*

Full particulars of other Drills free by post on application to

ATLAS IRON WORKS, EARLS COLNE, ESSEX.

65 *Hunt & Tawell's broadcasting barrow, 1875. This simple implement for sowing clover and grass seeds had been invented about 1830. The barrow, up to 17 feet long contained a revolving spindle with brushes that swept the seed out through apertures along the side. [35/15245]*

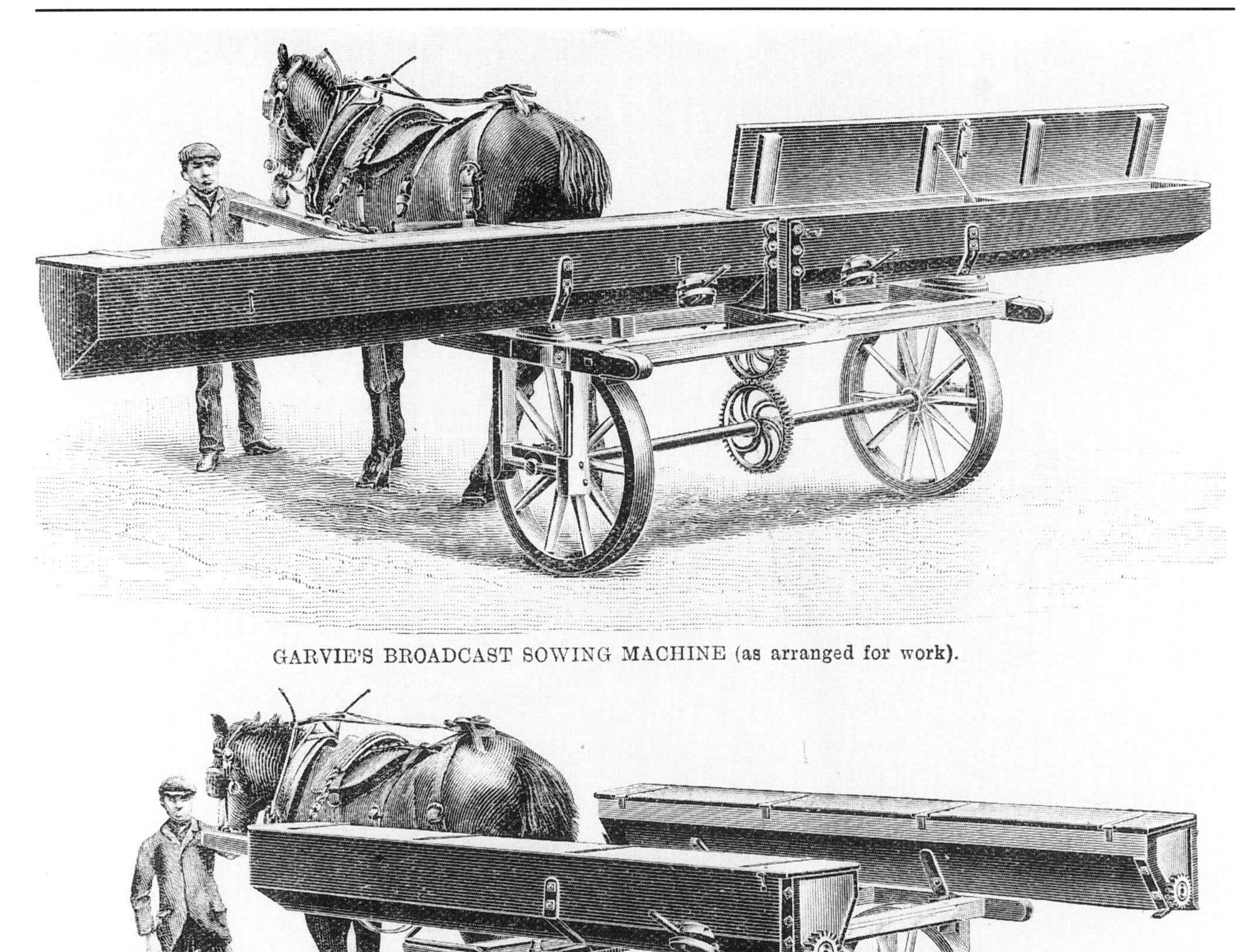

66 *Most broadcast barrows were hand held. The late nineteenth century produced some versions for horse haulage, such as this seed broadcaster and manure distributor by Robert Garvie & Sons of Aberdeen in 1894. [35/26393]*

RANSOMES, SIMS & JEFFERIES, LD., IPSWICH. 87

PATENT COMBINED STEEL CULTIVATOR, RIDGER AND HORSE HOE.

THE "TRIPLEX."

PATENT CULTIVATOR ARRANGED AS A RIDGER.

When required **as a Ridger** all the tines are taken out except three; on to these the ridging breasts are fixed and adjusted on the frame to suit the width required.

For Horse Hoeing the tines can be adjusted laterally on the bars to suit the ridges and are placed in three sets of three. The Machine will then clean three rows of roots or potatoes at one operation, varying from 24 in. to 30 in. in width.

The road wheels can be adjusted to suit the width of the ridges.

67 *Ransome's three-row ridger. Single-furrow ridging ploughs for potatoes and other root crops had been produced since early in the nineteenth century. Ransome's three-row version was an option on their patent 'Triplex' cultivator introduced about 1901, which could be used as a tined cultivator, a horse hoe or a ridger. [35/18422]*

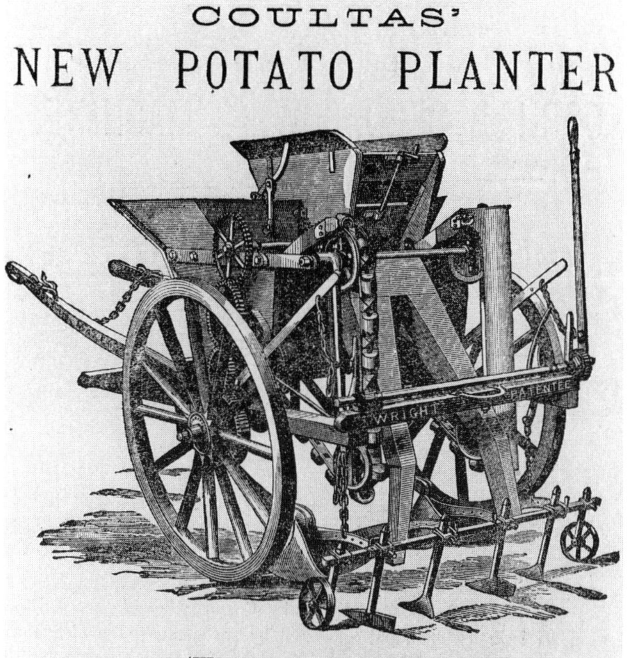

(WRIGHTS PATENT.)

THE ROYAL SOCIETY'S FIRST & ONLY PRIZE OF £15 AT BEDFORD, 1874.

This invaluable implement is constructed for Planting Potatoes in rows. It opens the Ridges, drops the Potatoes at equal distances, distributes any quantity of Artificial Manure, and covers them up at one operation, without in any way damaging the sets.

PRICES.	£	s.	d.
One-Row Planter, for dropping the Potato in the Ridge previously set up by a plough	18	0	0
Two-Row Planter (complete), as described, and according to engraving above, but without the Manure Apparatus	35	0	0
Two-Row Planter, complete in all details, as per description and engraving	45	0	0

Testimonials from large Potato Growers who have purchased the above can be had on application.

Small Seed Drills, Water Drills, Ridge Drills, Broadcast Corn Distributors, and other descriptions made to order.

SEE FULLY ILLUSTRATED CATALOGUES.

68 *Potato planter by Coultas of Grantham, 1875. The first patents for potato planting machines had been taken out in the 1850s, but it was not until the early 1870s that workable machines appeared on the market. This one had cups in a continuous chain to feed the potatoes from the box to the row. [35/26328]*

A JACK & SONS MAYBOLE

Top left **69** *The 'Caledonian' potato digger by Alexander Jack & Sons, Maybole. For harvesting potatoes in the early nineteenth century farmers had available potato raising ploughs, which had a pronged body to break up the ridges on which the potatoes were grown. Patents for digging machines were taken out in the 1850s, but it was not until the 1870s and 1880s that potato diggers, or 'spinners', were made in any numbers. They operated by a sideways revolving set of forks digging up the ridge to bring the potatoes to the surface. [35/6089]*

Left **70** *McCormick's reaper followed Patrick Bell's machine in having a revolving reel to draw the corn to a cutter bar of oscillating knives. Bell's had been designed with the bar at the front of the machine, so that it had to be* pushed *by the horses. McCormick placed it to the side so that the reaper could be* pulled. *The machine exhibited in 1851 had deposited the cut corn behind it. This illustration shows the reaper as modified from 1862 to throw the corn out to the side. [35/26325]*

First Prize Awarded this Year!

William Dray & Co. have again obtained the First Prize for their PATENT IMPROVED HUSSEY REAPING MACHINE, at the Meeting of the Bath and West of England Agricultural Society.

Prizes awarded in 1854 *by—*

The Royal Agricultural Society of England;
The Bath and West of England Agricultural Society;
The North Lancashire Agricultural Society; and
The Stirling Agricultural Society.

Numerous prizes have also been awarded to the same in previous years.

A Descriptive Catalogue may be had on application to the Manufacturers,

WILLIAM DRAY & Co.,

AGRICULTURAL ENGINEERS

SWAN LANE. LONDON.

71 *The Hussey reaper was a simpler machine, and being cheap to buy, at £20–£30, was quite popular in the 1850s and 1860s. However, with no means of holding the corn up to the knife it was difficult to use with a wet or damaged crop, while the manual raking was tiring and difficult to control. [35/6292]*

WOODCOTE, CARSHALTON, SURREY,
Sept. 10, 1874.

GENTLEMEN,—The two new Reaping Machines I had of you this season gave me the greatest satisfaction, and I think by degrees we approach a perfect machine.

The three great points to commend your machines are the lightness of draught, two horses being as easy as it is in pair-horse ploughing, the direct and level thrust of the knife, and the admirable contrivance for oiling the main bearings. By filling up every morning, they require no more attention all day, and for the cutting of 250 acres we did not use five pints of oil, and you will see by my account the breakages are *nil*. My entire crop was cut by two lads of fourteen or fifteen years of age, at threepence per acre, one going behind the other, and the average cut was one and a quarter acres each machine per hour all through, notwithstanding the expected difficult and expensive harvest. I never reaped it and saved it for less money; having it cut, tied, stocked, raked, carried, and stacked for *7s. 3d.* per acre. Never had a better crop, nor one of my labourers earned more money. Having this year again sold one of my machines, after finishing, I must ask you to book me one for next season; they are great favourites of all who have seen them.

Believe me yours faithfully,
JAMES ARNOT.

MANOR HOUSE, CASTLEACRE, BRANDON, NORFOLK,
October, 1874.

GENTLEMEN,—I cannot allow another season to pass without bearing testimony to the efficient working of the machine supplied to me last August.

I considered the one I had of you in 1871 was nearly perfection, and am more than pleased with the one you sent me this year. I cut wheat, barley, and beans with it this harvest, without a single breakage. I much like the seat for the driver, as I consider the horses ought only to drag the machine, without carrying the driver. The lubricators are an immense improvement, and the draught perfection. Wishing you every success,

Believe me,
THOS. MOORE HUDSON.

To Messrs. BURGESS & KEY.

BURGESS & KEY,
HOLBORN VIADUCT, LONDON, E.C.

Above **72** *The answer to the problem of delivering the cut corn in intermittent swathes convenient for binding into sheaves proved to be the arrangement of revolving rakes which swept the corn back from the cutter bar, around the wooden platform and out to the side. This self-raking, or 'sail' reaper, as it came to be known, was introduced in the 1860s by makers of McCormick-type machines, notably Samuelson of Banbury, Burgess & Key, and Hornsby. The illustration is of Burgess & Key's reaper advertised in 1874.* [35/6246]

Right **73** *Hornsby's 'Royal' self-raking reaper, 1875. The patent spring balance was a form of suspension intended to give the driver a smooth ride over the uneven field, despite the apparently precarious position of his seat.* [35/17598]

74 *The self-binding reaper which tied the cut corn into sheaves was introduced by McCormick's company in 1878. The principles of the design, with the corn being taken along a canvas conveyor to the binding mechanism at the side, were quickly established. The machine was not so quickly taken into use by the farmers, and it was the 1890s before sales began to build up. Hornsby again were among the major British manufacturers of binders. Their 12a 'open-back' binder was photographed in the works yard at Grantham in 1892.* [35/6531]

R. HORNSBY & SONS
GRANTHAM ENGLAND
HORNSBY.
853

The Patent "Albion" No. 1 Steel Binder.

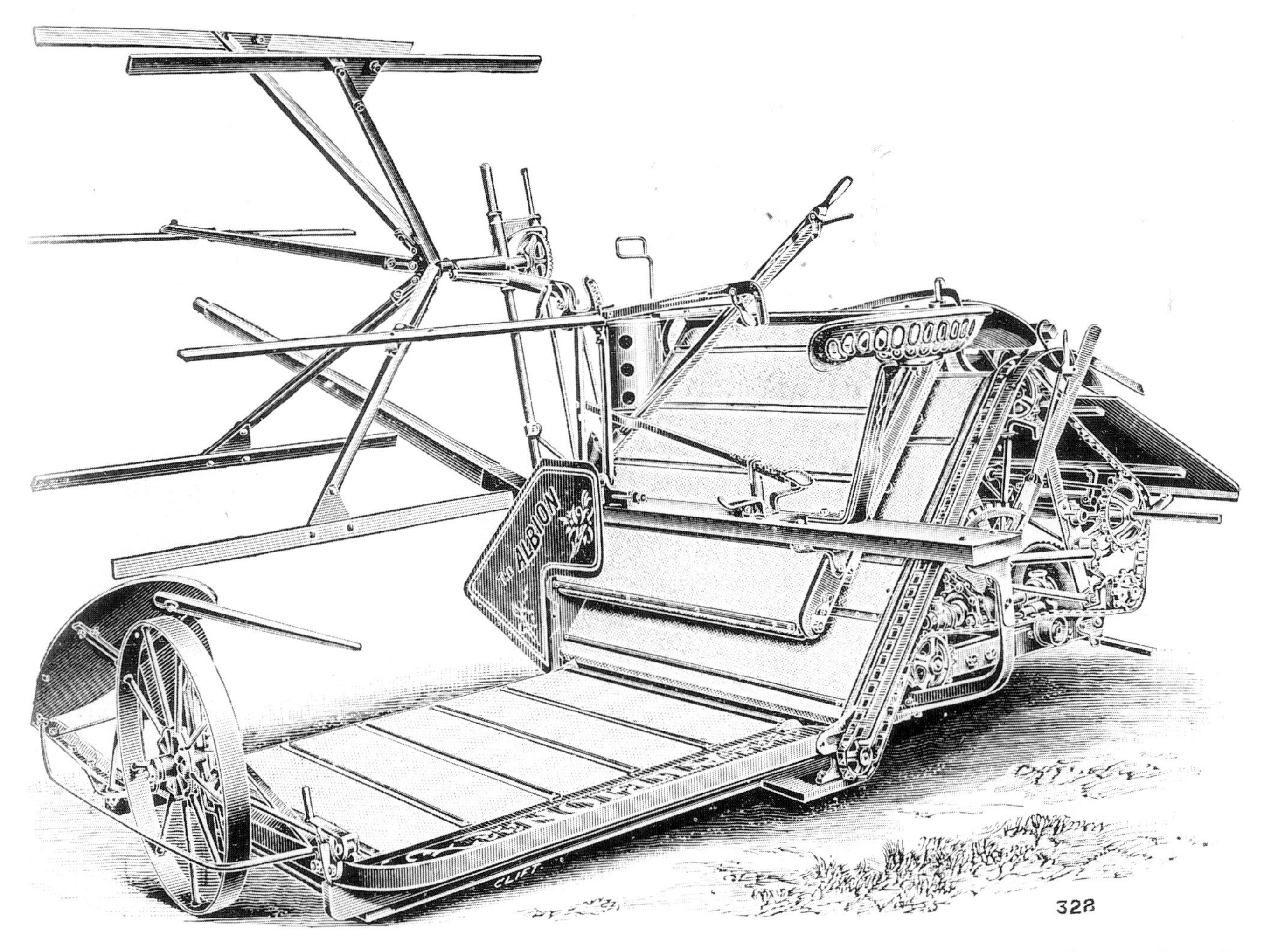

75 *Among the few new British names to achieve prominence in the world of agricultural machinery during the late nineteenth century was the Albion trade name of Harrison, McGregor & Guest of Leigh, Lancashire. This company became one of the top half dozen makers of harvesting machinery, carrying off prizes at international exhibitions, including the Universal Exhibition at Paris in 1889. The top illustration shows the No. 1 steel binder, illustrated in the firm's catalogue for 1900. The lower one is a restored No. 5, made in the 1930s for horse traction but later adapted for tractor haulage. [35/26275 & 60/14019]*

76 *The grass mower worked on the same principle as the corn reaper, but with smooth-edged, pointed knives rather than the serrated, triangular cutters that were best for the stiff corn crops. There was no need for the raking mechanisms to sweep the hay forward or away to the side. Bamford of Uttoxeter had been founded in the 1830s manufacturing stoves, but became one of the leading makers of hay harvesting machinery.* [35/26401]

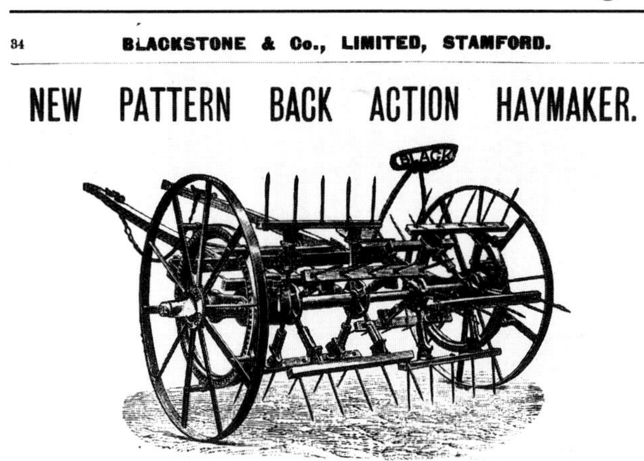

84 BLACKSTONE & Co., LIMITED, STAMFORD.

NEW PATTERN BACK ACTION HAYMAKER.

This machine has been introduced to meet the want of a good and cheap Haymaker in districts where single-action machines are preferred. It is very strong and durable, mounted on a solid steel axle, and is fitted with two springs to each fork. The height of forks is regulated by a very simple arrangement and without having to unscrew nuts. The cog gearing is securely fastened to the wheels and can be renewed at a small cost: this is a great improvement over other Haymakers having the spokes cast in the gearing. The teeth are behind the wheels when they turn the crop, consequently the weight of the machine is not on the hay when the teeth come in contact with it. This renders the machine LIGHT IN DRAUGHT, and it can be worked in heaviest crop by a light horse.

	Extreme Width.	Width over Tires.	Width over Flyers.	Height of Wheels.
No. 4, with 4 levers to each flyer	7ft. 10in.	6ft. 8in.	5ft. 4in.	4ft. 0in.
No. 5, with 5 „ „	7ft. 10in.	6ft. 8in.	5ft. 4in.	4ft. 0in
No. 8, with 6 „ „	8ft. 4in.	7ft. 3in.	5ft. 11in.	4ft. 6in.

Driver's seat, extra.

The No. 8 is a very powerful Machine, and sufficiently wide to take two swaths as cut by an ordinary Mowing Machine.

Our Patent Haymaking Machines WON

All the FIRST PRIZES of the Royal Society of England for 10 years;
All the FIRST PRIZES of the Royal Society of Ireland for 10 years;
All the FIRST PRIZES of the Royal Society of Scotland for 10 years; and the
FIRST PRIZE of £20 and the SECOND PRIZE of £10
At the Last Trials of the Royal Agricultural Society of England.

FOR PRICES, SEE PAGE 6 IN LIST AT END OF CATALOGUE.

Left **77** *Back action haymaker by Blackstone of Stamford. Mechanical cutting of the hay, leaving the crop in continuous swathes increased the demand for hay making machinery. [35/15251]*

78 *Some new machines for turning the mown hay were produced in this period. Jarmain's patent swath turner was made by Ransome. It worked with two sets of revolving blades positioned to turn the swath onto a neighbouring dry strip of ground. This one-horse machine could, according to the makers, turn 20–30 acres a day. [35/8465]*

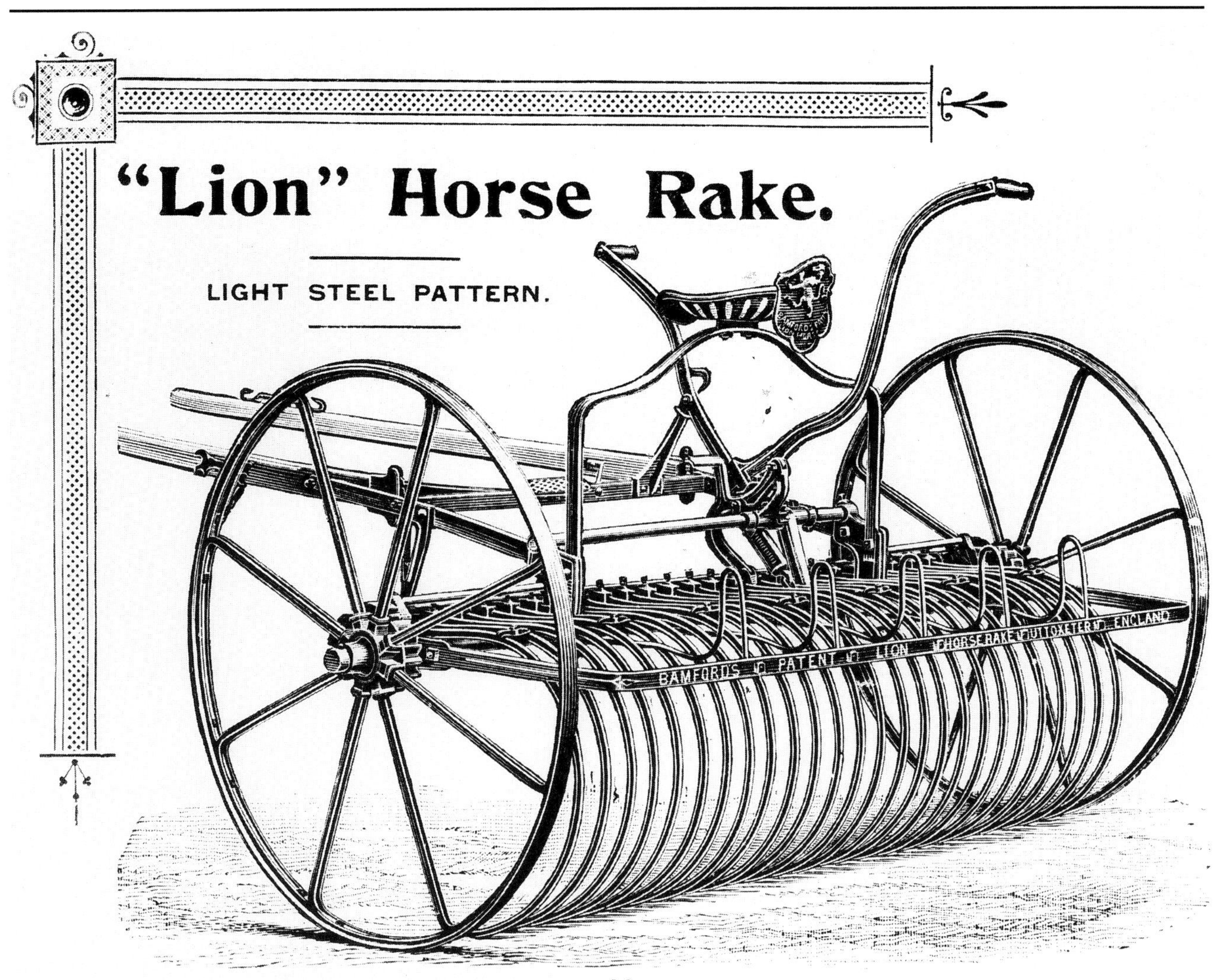

79 *A Bamford rake, the 'Lion' of 1897 (top). The hay rake was another implement that achieved greater use with mechanized cutting of the crop, although the employment of gangs of labourers wielding wooden rakes remained common until the First World War. Horse rakes were being made in the 1830s and 1840s, but it was not until the 1850s and 1860s that rakes with efficient mechanisms for lifting the teeth were developed. Few agricultural implements were imported from France, but the bottom rake was brought over about 1900 and used on a farm in Somerset. It was made by the firm of Pilter, a leading agricultural engineering concern in France that acted as agents for a number of British manufacturers, including Garrett of Leiston and Ivel tractors. [35/26400 & 60/14708]*

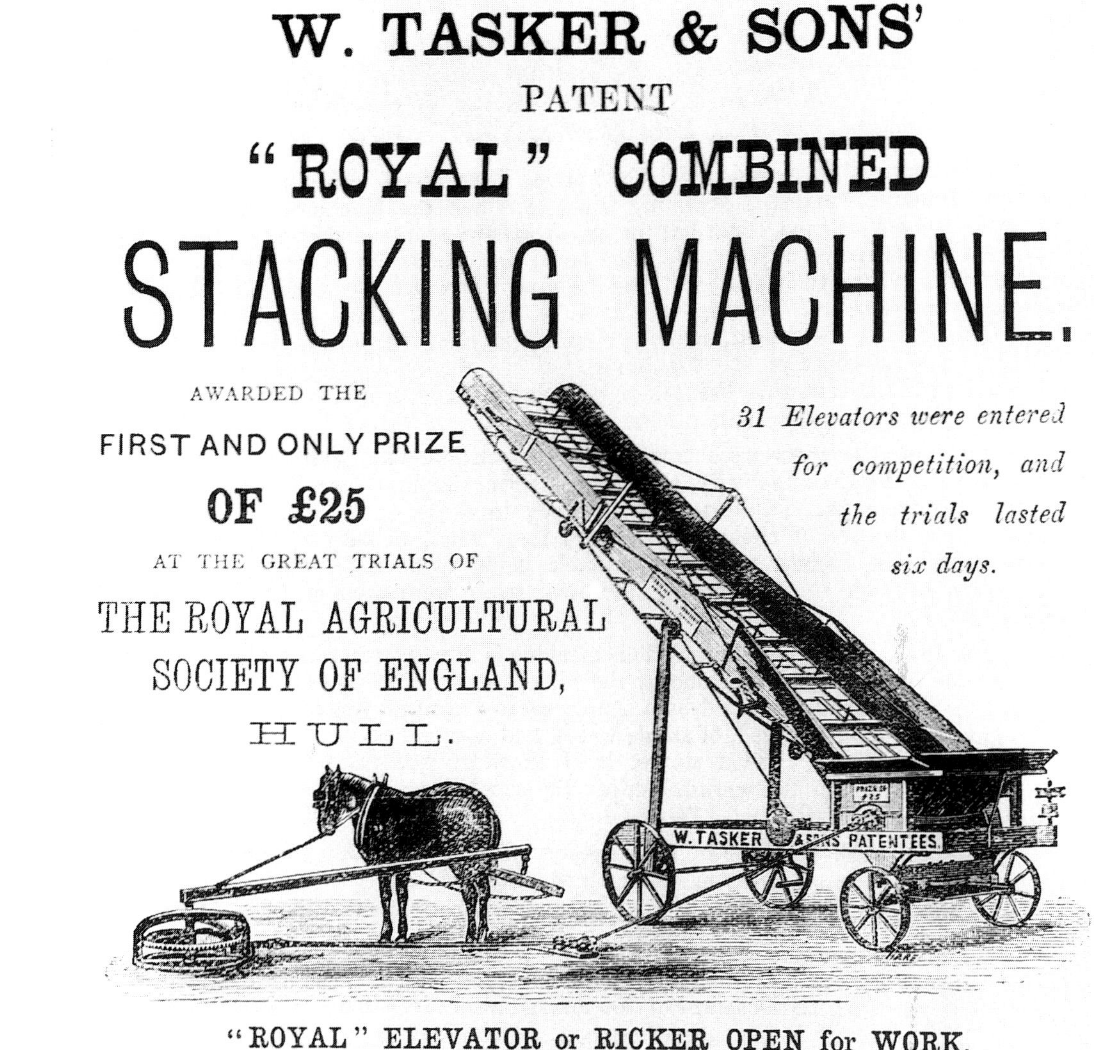

"ROYAL" ELEVATOR or RICKER OPEN for WORK.

No. 10 "Royal" Elevator, to deliver 26 feet from the ground ...	£46 10 0
If fitted to work with a Thrashing Machine, to deliver straight, extra	2 0 0
If fitted to work with a Thrashing Machine, to deliver at any angle, extra	3 10 0
Light double-speeded portable one-horse gear to drive the Elevator in hay-time and harvest when rickmaking	8 0 0

When travelling the gear work is attached behind the Elevator.

For working in confined Stack-yards a simple but substantial arrangement is made for using only one coupling bar from the horse-gear to the elevator. Extra charge for same 40s.

A slight modification of the "Royal" Stacker renders it admirably adapted for Stacking Bark. Particulars on application.

The "Royal" Elevator or Stacker delivers to a height of 26 feet, and it was proved conclusively at the Hull Trials that it elevated the produce a greater number of feet than any other brought for trial. At these trials W. Tasker & Sons met all the leading makers, and, after the most thorough and searching trials ever made, were awarded the whole amount offered for competition, they have therefore not thought it worth while to compete at any local trials.

WATERLOO IRON WORKS, ANDOVER, HANTS.

Left **80** *Elevators for building hay and straw ricks began to appear in the 1860s, and ten years later several of efficient design and reasonable price, for horse or steam power, were on the market. Tasker of Andover offered the 'Royal' elevator in 1877.* [35/26378]

Right **81** *The hay pole, a simple form of crane worked by hand or by horse gear, was another means of building the stack. This photograph shows it in use in the early years of the present century on a farm in Northumberland, a region where it found particular favour.* [35/7316]

82 *The hay press was introduced in the 1880s. A hand-operated lever press advertised in 1895 by W. J. & C. T. Burgess of Brentwood, Essex, successor company to Burgess & Key.* [35/26393]

IMPROVED PORTABLE LEVER PRESS

These Lever Presses, being Strong, Portable, Compact, Convenient, and Cheap, have found their way readily into the Colonies and other Countries where Hay, Straw, Cotton, Husks, Hides, Rags, Paper, Skins, Hemp, Moss, Wool, and numerous other kinds of merchandise require to be put into neat and compact bales for transportation; they are easily worked by two men. The inside measurement of the standard size Press, when closed up is 45 X 26 X 26 ins.; slight variations from these sizes are made without extra charge. For hay, four or even more trusses may easily be compressed into the above space, and for cotton, 320 lbs. can be pressed into a space of 48X28X28 ins.

This Press is an improvement upon the American Press in which the chain was drawn up by means of small teeth between the ratchet wheels, these teeth were required exactly to fit the links of the chain, consequently a very little wear of the chain or of the teeth caused the chain to slip over the teeth and thus render the Press useless. In this improved Press, as shown in the drawing, the chain is coiled up upon a grooved cone, the end being fastened to the large end of the cone; the winding commences at the large end, and therefore winds up much more chain at first for each depression of the lever than it does at the last, thus increasing the power in proportion as the resistance of the bale becomes greater. The end of the chain being fastened securely to the cone, there is no possibility of any slip occurring. This Press when packed for shipment, measures 56 cubic feet. Gross weight, 12 cwt.

Price £22, delivered to the Docks in London.

RICHARD GARRETT & SONS,

LEISTON WORKS, SAXMUNDHAM, SUFFOLK,

BEG RESPECTFULLY TO CALL ATTENTION TO THEIR

NEWLY-INVENTED 'FIXED' AND 'PORTABLE' COMBINED

FINISHING, THRASHING, AND DRESSING MACHINES.

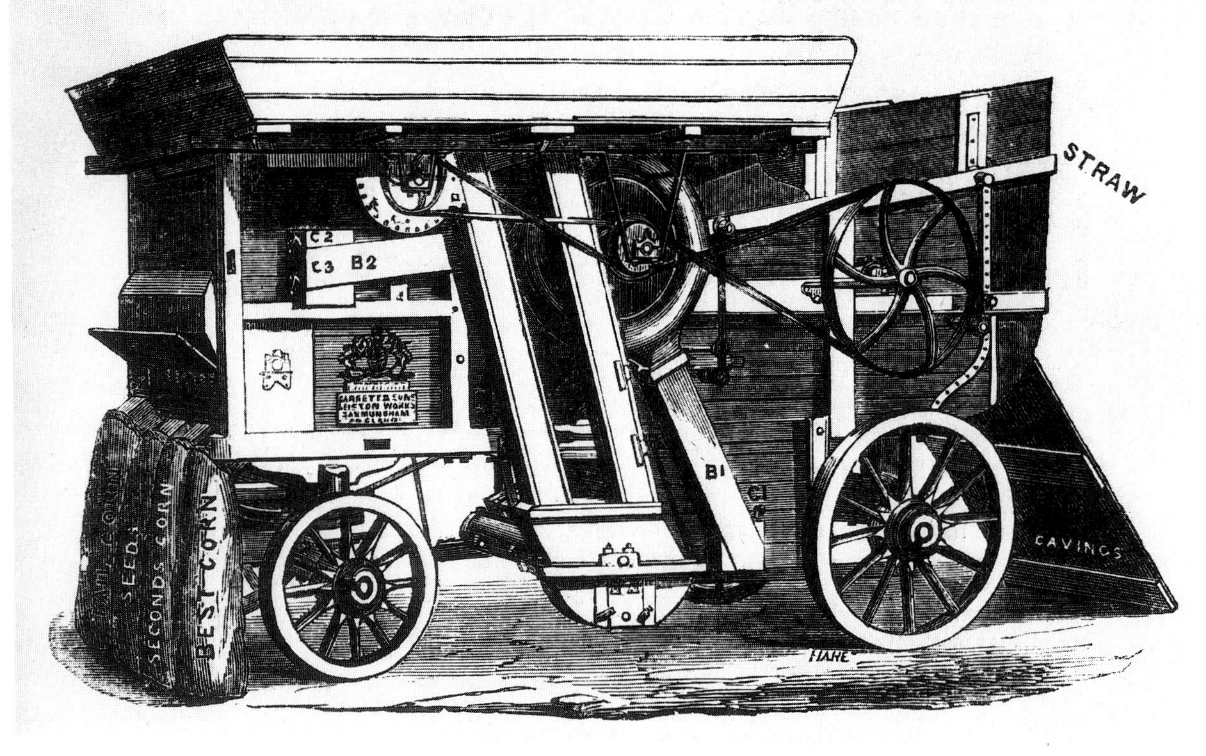

83 *Garrett's combined threshing, finishing and dressing machine of 1859. By this time combined machines were already becoming standard equipment for large farms and for the contractors. [60/9322]*

Top right 84 *The basic sequence of operations in the combined threshing machine had been settled by the late 1850s. Subsequent development was to make the machine both more sophisticated and complicated, as this diagram of a machine made by P. & H. P. Gibbons of Wantage in 1879 shows. The machine was equipped at the top with a self feeder, a recently-introduced safety device which avoided the men having to work by the open mouth of the threshing machine. This one was a double blast machine, there being two winnowing blowers (at O and V) before the finished grain finally emerged into the sacks. [35/10336]*

85 *Ransomes, Sims & Jefferies class A threshing machine, 1902, one of the company's large double blast machines. It was available with the drum sizes from 48–60 inches, and was made, with modifications, until threshing machines went out of production in the 1950s. [35/7953]*

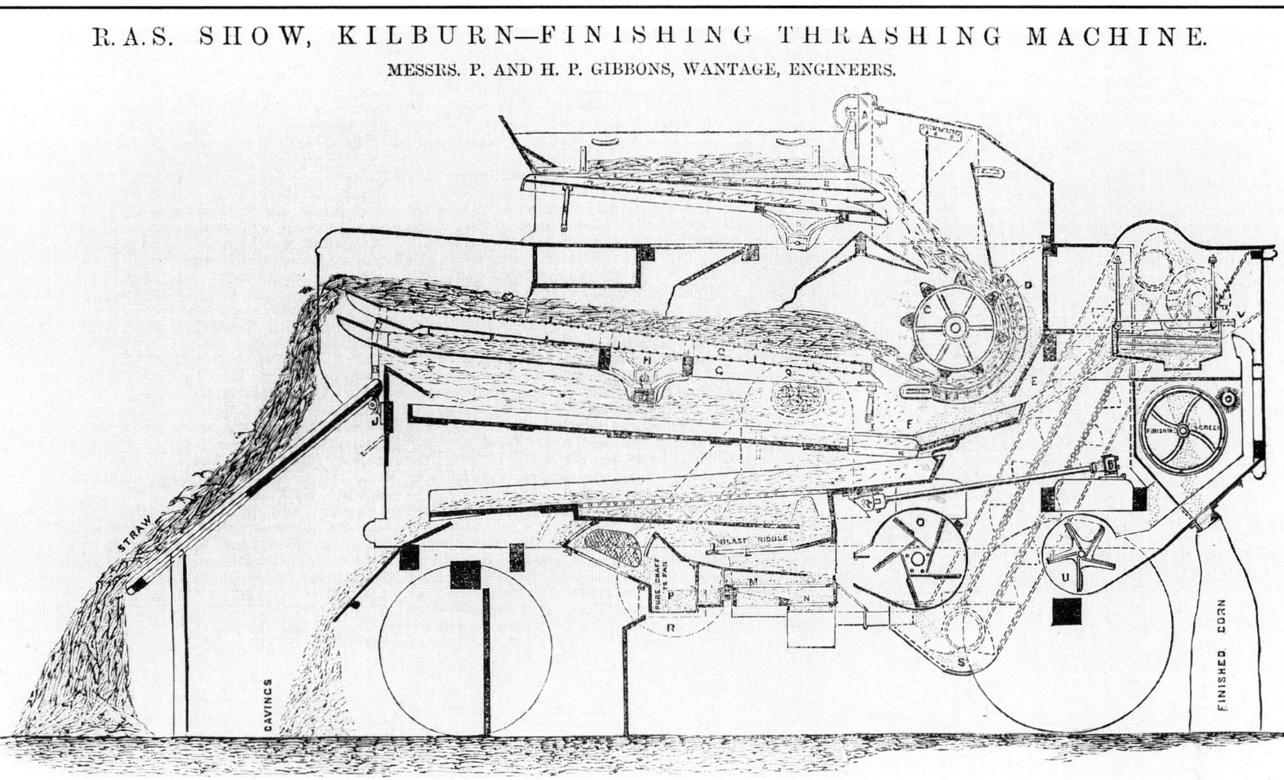

R.A.S. SHOW, KILBURN—FINISHING THRASHING MACHINE.

MESSRS. P. AND H. P. GIBBONS, WANTAGE, ENGINEERS.

Top **86** *Hornsby's straw trusser, 1887. Machines operated in conjunction with the threshing machine came into use mainly from the 1880s. One of these was the straw trusser, attached to the end of the threshing machine to bind the loose straw.* [35/26394]

87 *Straw-baling presses attached to the thresher and driven by the steam engine were also available by the end of the 1880s. This is Howard's baler of 1889.* [35/11633]

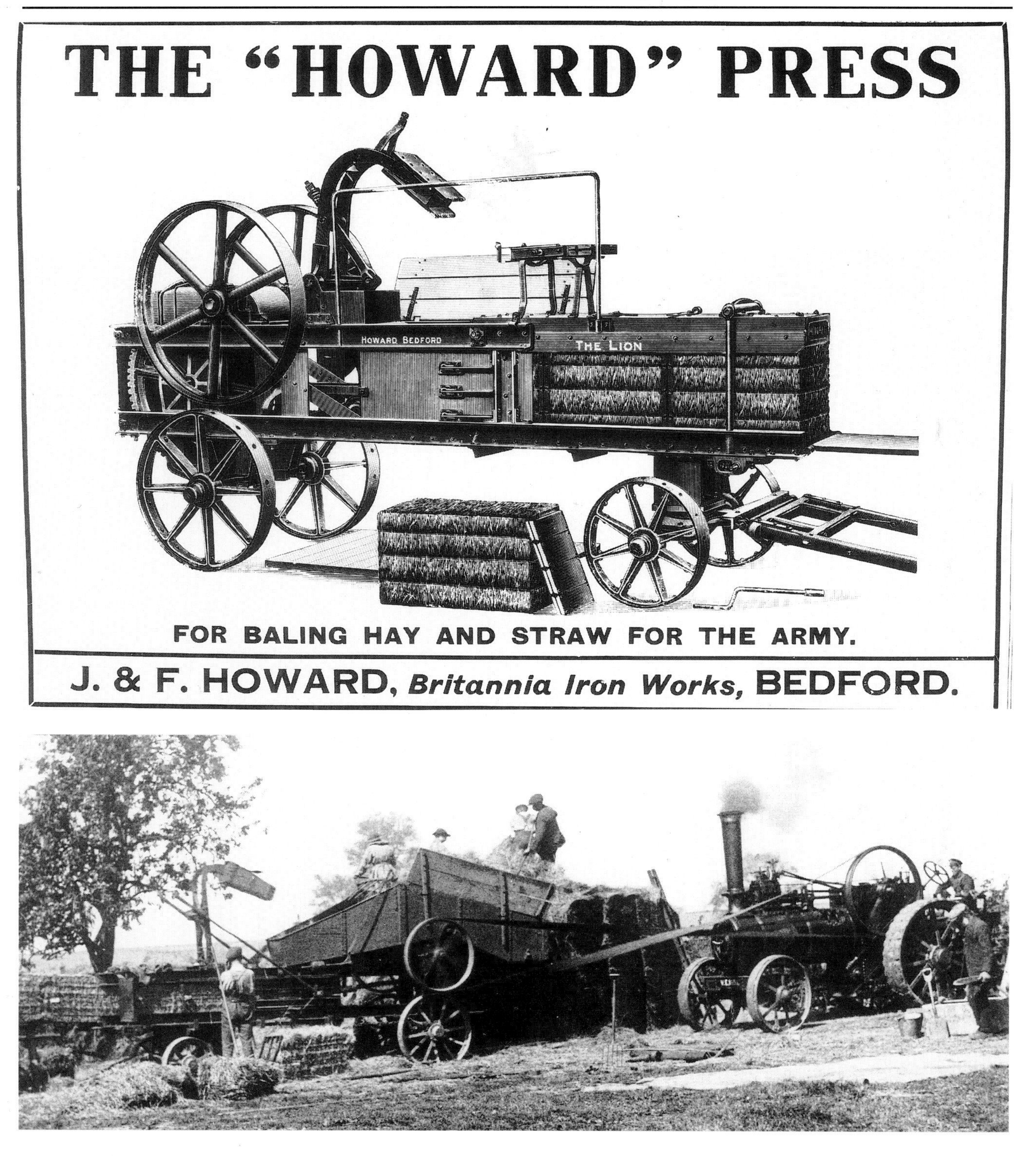

Top **88** *The Howard 'Lion' baling press with a reciprocating ram arm was of a type introduced in the early years of the present century, and was becoming standard by the First World War when this was advertised. As well as operating as a straw baler alongside a threshing machine, it could be worked independently as a hay press. [35/6827]*

89 *Ruston straw baler powered by a Ransome traction engine during the First World War. [35/26395]*

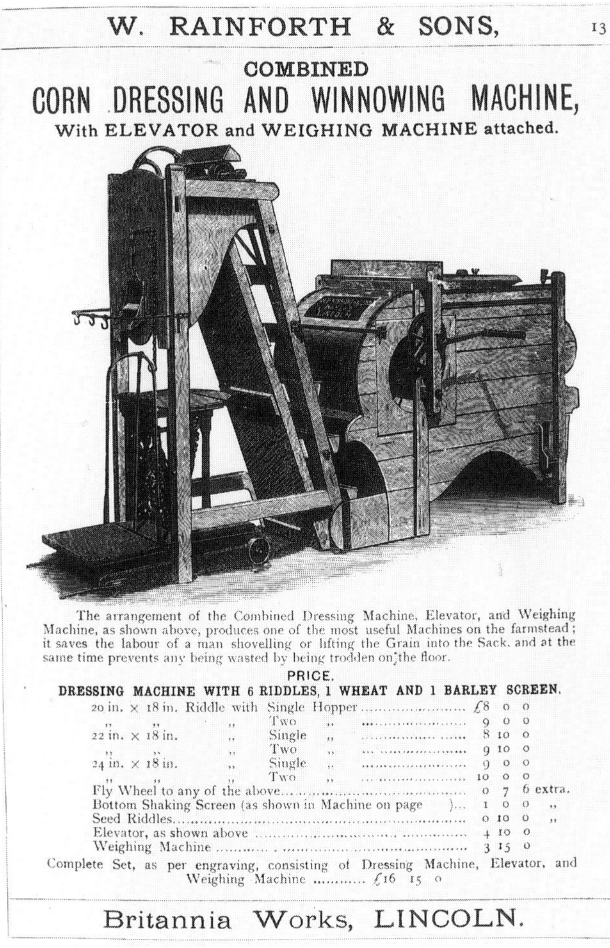

W. RAINFORTH & SONS, 13

COMBINED

CORN DRESSING AND WINNOWING MACHINE,

With ELEVATOR and WEIGHING MACHINE attached.

The arrangement of the Combined Dressing Machine, Elevator, and Weighing Machine, as shown above, produces one of the most useful Machines on the farmstead; it saves the labour of a man shovelling or lifting the Grain into the Sack, and at the same time prevents any being wasted by being trodden on the floor.

PRICE.

DRESSING MACHINE WITH 6 RIDDLES, 1 WHEAT AND 1 BARLEY SCREEN.

	£	s.	d.	
20 in. × 18 in. Riddle with Single Hopper	£8	0	0	
,, ,, ,, Two ,,	9	0	0	
22 in. × 18 in. ,, Single ,,	8	10	0	
,, ,, ,, Two ,,	9	10	0	
24 in. × 18 in. ,, Single ,,	9	0	0	
,, ,, ,, Two ,,	10	0	0	
Fly Wheel to any of the above	0	7	6	extra.
Bottom Shaking Screen (as shown in Machine on page)	1	0	0	,,
Seed Riddles	0	10	0	,,
Elevator, as shown above	4	10	0	
Weighing Machine	3	15	0	

Complete Set, as per engraving, consisting of Dressing Machine, Elevator, and Weighing Machine £16 15 0

Britannia Works, LINCOLN.

Above **90** *A Ransome traction engine driving a 54-inch threshing machine and chaff cutter, early 1900s.* [35/14938]

91 *Despite the advent of the large threshing and finishing machines, there was still a demand for the smaller items of barn machinery powered by hand or by horse gear. Rainforth's dressing and winnowing machine illustrated in a catalogue of the 1880s together with an elevator and bagging apparatus.* [35/6855]

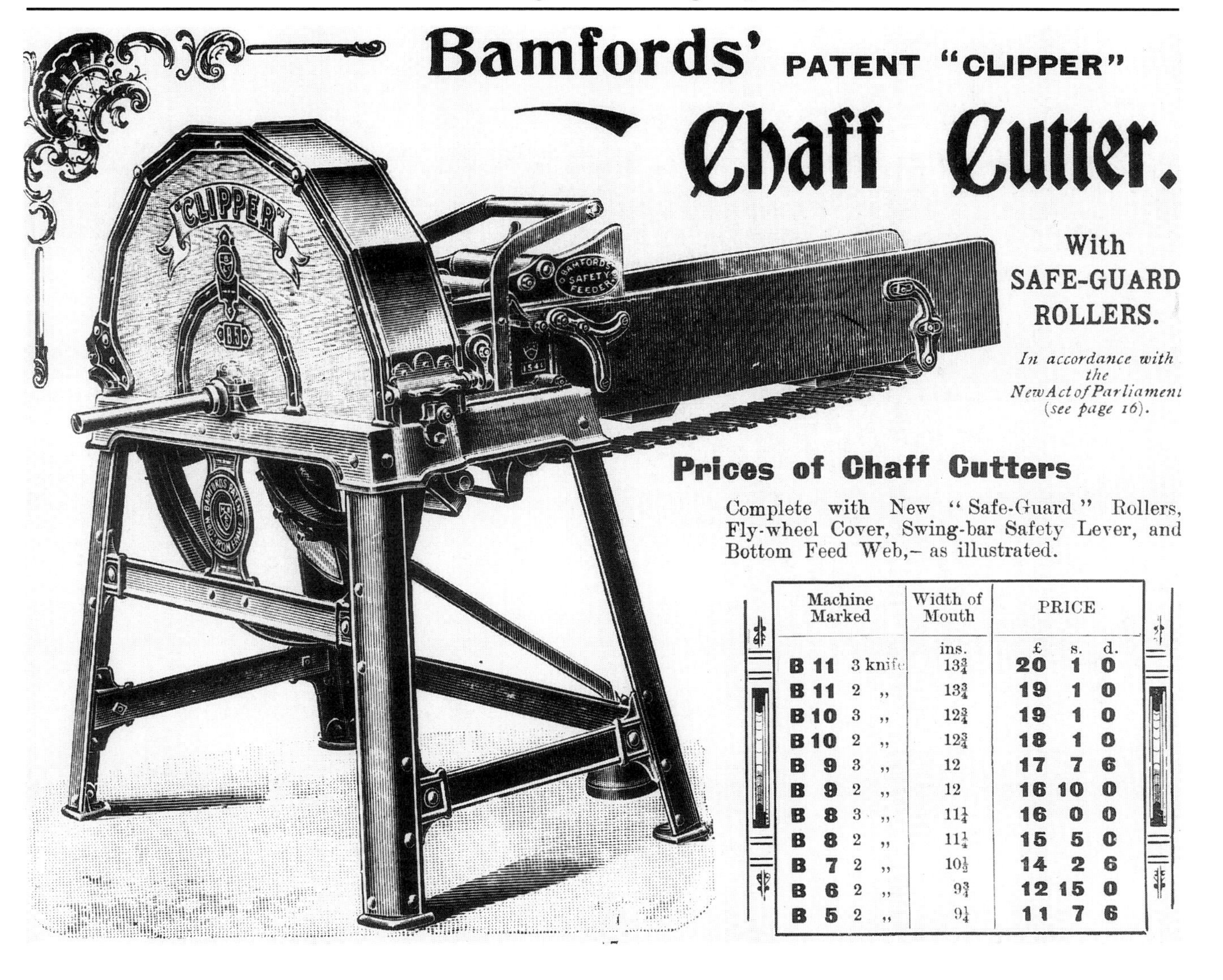

Machine Marked		Width of Mouth	PRICE		
		ins.	£	s.	d.
B 11	3 knife	13¾	20	1	0
B 11	2 ,,	13¾	19	1	0
B 10	3 ,,	12¾	19	1	0
B 10	2 ,,	12¾	18	1	0
B 9	3 ,,	12	17	7	6
B 9	2 ,,	12	16	10	0
B 8	3 ,,	11¼	16	0	0
B 8	2 ,,	11¼	15	5	0
B 7	2 ,,	10½	14	2	6
B 6	2 ,,	9¾	12	15	0
B 5	2 ,,	9¼	11	7	6

92 *Chaff cutters were produced to this pattern by several manufacturers. The design changed little over the years, the main improvements being to the rollers, and the fitting of safety guards for the rollers and the cutter. Bamfords' 'Clipper' was featured in the firm's catalogue for 1900. [35/17783]*

Bamfords' "PATENT RAPID" Grinding Mills

EXCEL all others for Fine Grinding, Simplicity, Durability, and supply the long-felt want of a simple strong mill, with all parts freely accessible, and so constructed that an intelligent labourer can easily manipulate, or readily replace any duplicate part.

The Grinding Surfaces being larger than in other Mills, the **"RAPID"** gets through **more work,** does it **quicker,** and **lasts longer.**

When the plates are worn on one side a labourer can reverse them.

THE MILL IS THEN EQUAL TO NEW!

"SIMPLEX" Detachable Split Pulleys. The "Rapid" Mills are fitted with these Pulleys, which are readily transferable if desired to drive from the end of the Main Shaft outside the Mill Frame.

LATEST RESULTS—Continued Success ! !
Highest Prize Silver Medals at the Royal Danish Trials, Randers; at the Warwickshire Show; and at the London Dairy Show.

TWO FIRST PRIZE SILVER MEDALS at the Durban Society's Trials (South Africa).

Mill Marked.	Price.	Power Required.	Speed Revolutions per minute.		Capacity in Bushels per hour.		Reversible Grinding Plates per pair.
			Horse Gear.	Steam, &c.	Fine Meal.	Splitting or Kibbling	
No. 0—h.	£4 5	For Hand	—	—	—	6	7/6
0 p.	4 5	Pony Gear	400	—	3	10	7/6
1—H	7 10	1 Horse	400	600	4 to 6	16 to 20	10/-
1—H M	8 0	1 ,,	—	—	4 to 6	16 to 20	10/-
1—H.G.	9 0	1 ,,	—	—	4 to 6	16 to 20	10/-
No 1	9 0	1 or 2 Horse	350	400 to 500	5 to 8	25	12/6
No. 2	11 10	2 or 3 ,,	350	400 to 500	8 to 10	35	15/-
No. 3	16 10	3 H.P.	—	400 to 500	14 to 16	45	17/-
No. 4	19 10	4 H.P.	–	400 to 500	16 to 20	60	20/-
No. 6	27 10	6 H.P.	—	450 to 500	20 to 30	80 to 100	30/-

A MONTH'S FREE TRIAL.
EVERY MACHINE GUARANTEED TO GIVE SATISFACTION.

JULY, 1898—FIRST PRIZE at NATAL (South Africa), to the "RAPID" GRINDING MILL.

93 *Another of Bamfords' barn machines was the grinding mill for crushing oats, beans and peas. Simple mills for crushing grain for livestock feed had been available since the mid-eighteenth century. By the late nineteenth century this was the most usual design, for working by hand or horse gear.* [35/17782]

Top right 94 *Bentall of Maldon was another of the prominent makers of barn machinery, of which their OCK cake breaker was an example of the 1900s.* [35/26403]

Bottom right 95 *Steam engineering: the boiler shop at Marshall's Britannia Works in Gainsborough, late nineteenth century.* [35/18275]

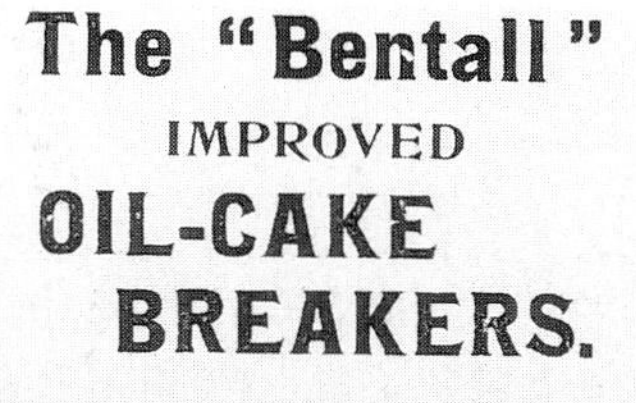

The "Bentall" IMPROVED OIL-CAKE BREAKERS.

MARK.	in.	£	s.	d.
OCK	Mouth $12\frac{1}{2}$	2	17	6
OCH	" 12	3	3	0
OCM	" 12 as illustrated	3	6	0
OCP	" 14	3	10	0
OCO	" 16 with 2 handles	3	16	0
OCR	" 18 " 2 "	4	1	6
OCC	" 12 " 2 "	5	10	0
OCE	" 12 " 2 "	6	12	0

These machines, by means of the eccentric, can be regulated to break seven different sizes.

In the Machines marked OCM, OCP, OCO, OCR and OCS, the gear wheels are completely enclosed.

The OCR is a new machine for breaking extra wide cake.

The above machines, and also the OCK, have improved hoppers, one side of which can be set back, to receive large quantities of broken pieces.

The OCH is an extra strong machine.

The OCC and OCE Mills are fitted with extra rollers and concaves so as to reduce the cake sufficiently fine for lambs and calves, the rollers and concaves can be thrown out of use when lump cake only is required.

The OCS is a strong machine for power, but it is geared so that it can be easily used by hand. It has **two pairs of rollers** so that the cake can be reduced sufficiently fine for lambs or calves.

All the bearings are renewable, and being fixed to the Machine by bolts, can easily be replaced when worn. All the Gearing is covered with ornamental cast-iron covers.

Mark OCS OIL-CAKE MILL, Mouth $16\frac{1}{2}$-in. wide, £9 9 0

Extras :—**Pulleys.** For prices see page 39.
Belt Shifter for fast and loose pulleys, see page 30, price 20/-.

Power, $1\frac{1}{2}$ to 2 B.H.P. Speed, 160 Revolutions per minute. Capacity, 30 to 35 cwt. per hour.

It will crush locust Beans for feeding Horses, etc.

The OCS if specially ordered can be supplied with rollers made of **steel ratchets** for exceptionally hard work, the extra cost for which is 65/-.

FIVE
The Twentieth Century: 1914–1945

The greatest change in farming technology during the twentieth century was the application of the internal combustion engine to agriculture. Like the steam engine, it was taken up first and most quickly as a stationary engine to drive chaff cutters, seed dressers, mills and other barn machinery. In this form it was already becoming established in farm work by 1900. The first engines on the market had been gas engines, powered by coal gas, and based closely upon Otto's prototype four-stroke engine of 1876. Gas engines retained a place in agricultural work, but the petrol and oil engines introduced during the 1890s and 1900s, with their more convenient source of fuel, became the farmers' favourites. In 1925 the estimate was that 1125 gas engines were in use on the farms of England and Wales, but more than 56,000 petrol and oil engines. Numbers of the latter rose steadily: 131,000 in 1942, and a peak of 221,000 in 1952. The engines farmers bought were commonly in the range $1\frac{1}{2}$–3 horsepower, although on larger farms those up to 5–7 horsepower were also popular.

Ultimately it was the tractor that had the most dramatic effect on the mechanization of farming. It was also to lead to a stronger presence of North American companies in the British farm implements market. The first tractor was American, made by the Charter Engine Company of Chicago in 1889, and several firms had by 1914 become established as tractor makers in the United States, among them Hart-Parr, Wallis, and Moline. However, development was active on this side of the Atlantic also, some indeed, of great importance. For while most manufacturers were making what were in effect traction engines with an internal combustion engine replacing the steam engine, in 1902 Dan Albone, of Biggleswade, produced the first really light petrol-engined vehicle intended for all work in and around the farm, in the fields and on the road. Albone held several successful demonstrations of his 'Ivel' tractor's prowess in hauling ploughs, mowers and binders on farms near Biggleswade. The Ivel had a great influence on the design of tractors on both sides of the Atlantic. It was also a reasonable success commercially: about 900 Ivel tractors were built between 1902 and 1921, when the firm closed down. Other British tractor makers were becoming established during the years down to 1914. The Saunderson, the product of another singularly inventive engineer, H. P. Saunderson, was the best-selling British make. The long-established firm Marshall of Gainsborough was another leading company. All three found the home market extremely limited, and derived most of their business from exports.

The First World War saw the first significant increase in the number of tractors on British farms, especially after the government took the supply of agricultural machinery in hand as part of the food production campaign. Substantial orders were placed for Saunderson and other British makes, but with military work having priority in the engineering industry, the government turned to companies in the United States to supply additional tractors. The bulk of the 15,000 tractors exported from America in 1917 were destined for this country and makes rarely seen in Britain before 1914 were becoming more familiar; Titan, Mogul, Overtime, Moline. Above all there was the Fordson. The government ordered 6,000 Fordsons in 1917 despite the fact that the tractor had not then gone into full production, but was still being demonstrated as a prototype. It proved to be an inspired choice, for the tractor was reliable and most popular with the farmers. Its novel frameless construction was also highly influential, establishing the lines upon which the design of tractors has been based ever since.

Its successful introduction during the final year of the war helped establish the Fordson as the dominant tractor in British farming throughout the period to 1945. It easily saw off the British competition, which for a few years was quite variegated. Besides the established firms, Saunderson and Marshall, several companies entered the tractor market between 1918 and 1920, Austin, Blackstone, Glasgow and British Wallis being among the more prominent. Some of these tractors were quite respectable machines, but they were poor value for money beside the Fordson. In 1921 when the British Wallis cost £525 and the Austin £360 the Fordson could be bought for £260. As agricultural markets became depressed in the 1920s the British tractor makers withdrew, although Marshall came back in the 1930s with new diesel-powered tractors. The large American companies, such as International Harvester, with their popular 10–20 tractor, Case and Caterpillar, retained a stronger presence, but not enough to shake Fordson's dominance. During the 1940s, Fordsons, made at Dagenham, accounted for more than 90 per cent of the tractors made in Britain, and their share of the total of tractors in use in the country was not much less.

The advance of the tractor was slow during the 1920s and 1930s. There were 46,500 tractors on the farms of England and Wales in 1937, a total that seems grand enough until it is compared with the 549,000 horses that were then at work in agriculture. The continued balance in favour of horse power rather than brake horsepower is one reason why tractors had less effect on the design and production of other implements than they might have done. Manufacturers were quick to offer implements for use with tractors. For example, Ransome's first tractor plough appeared in 1907 and during the succeeding years dozens of other implements became available. To a great extent, however, the implements for tractors were simply larger versions of horse-drawn implements: seed drills that could sow 20 rows of corn instead of 12 or 14, rollers and harrows that could be used in gangs of three. The tractor was being used as often as not simply as a substitute for the horse in pulling implements across the field, rather than as a source of power to drive machinery. The parsimony of farmers through the depressed years of the 1920s and 1930s in replacing the horse shafts with a drawbar for a tractor on existing implements did not encourage the manufacturers to produce tractor-powered machines.

The means for using the power of a tractor's engine to drive other implements were readily available. Power take off, slow-running shafting from the engine, which could be coupled to trailed implements, had been invented in France before the First World War. After the war it was championed mainly by the International Harvester Company on whose tractors power take off was fitted as standard equipment. The same company was responsible for introducing the row-crop tractor, a tractor with narrow wheels and the body raised well clear of the ground to allow tools to be arranged underneath as well as in front and behind the tractor. The tools were mounted on tool bars connected to the power take off by means of which they could be set into and out of the ground. Row-crop tractors were enormously popular in America, but considerably less so in this country before the Second World War; their main use was in potato and sugar beet fields.

In the late 1930s a yet more important development in power transmission by tractors became commercially available. Harry Ferguson demonstrated his system of hydraulic lift and three-point linkage in 1934–5. Through it the driver had more complete control over the operation of his implements. Ploughs no longer needed ground wheels, for not only was the plough lifted in and out of the ground by power from the tractor's engine, but the hydraulic circuit could be used to control the draught and the depth of the furrow. All other implements could be controlled in a similar fashion. Tractors incorporating Ferguson's hydraulic system were made by the David Brown Gear company of Huddersfield. Between 1936 and 1939 about 2,000 of these Ferguson-Brown tractors were made. A minute total beside the figure of 20,000 tractors a year which Ford's Dagenham plant was then producing.

Of other developments in farm mechanization during this period, many were connected with harvesting. For haymaking there were side delivery rakes, often combined rakes and swath turners. They were introduced in 1903, but found little market before 1914.

The hay loader was an even older invention, introduced from America in the 1870s, but again it was rarely used before the 1930s. The pick up baler, another machine of American origin, was beginning to be used in this country by the end of the 1930s.

The most important new harvesting machine was the combine harvester. It had been developed into a practical and commercial machine in the 1860s and 1870s in America, and by the 1890s was in common use there. But it was not until 1928 that the combine harvester was first used in Britain. It was taken up but slowly; by 1939 there were still fewer than 100 combine harvesters at work in this country. The main reason for this was that combine harvesters at this time were machines designed for work in the prairies and were thus not economical to operate in fields of less than 250 acres, which in British conditions was a large field.

The Second World War began to change that. The shortage and expense of labour brought down the minimum acreage at which the combine harvester became a sound proposition. In 1942 there were 1,000 combine harvesters in Great Britain; and 3,460 by 1946. There was a similar story with all other types of machinery and farm implements. The numbers of tractors rose rapidly during the war from 56,200 in 1939 to 203,400 in 1946, potato spinners and diggers increased in number from 37,980 in 1942 to 64,620 in 1946, and in general all types of implement for use with tractors were more widely used. A more thorough-going food production policy which substantially increased the acreage of arable land, and of cereal crops in particular, meant that the Second World War had a far more profound effect on the mechanization of farming than the First World War had done. It was claimed at the time that British farming at the end of the war in 1945 was the most highly mechanized in the world. Maybe that was so, but it has turned out to be even more true that agriculture then stood on the threshold of some of the most revolutionary changes in its history.

96 *The Ivel tractor, an illustration of the original model from the* Implement & Machinery Review *for September 1902. A modified version of the tractor with a more powerful engine of 14 bhp in place of the 8 hp of the original was exhibited at the Royal Agricultural Society's show in 1903 and again in 1904, when it was awarded a silver medal. By that time the tractor had achieved considerable technical success. It could be purchased new in 1903 for £300. [35/8084]*

97 *Saunderson Universal tractor, 1919. H. P. Saunderson made his first tractors in 1896–7, but it was not until 1904 that the first model to be available commercially appeared. The tractor went through several transformations of design during the next few years, some versions unconventional both mechanically and in appearance. The post-war model G, shown here, was of more straightforward design, a four-wheeled tractor with a two-cylinder engine developing 23–25 bhp. Although it had achieved reasonable commercial success up to 1920, the Saunderson company, of Bedford, fell victim to the depression in agriculture of the 1920s. [35/26278]*

Above **98** *The International Harvester Titan 10–20 was one of the most popular of the American tractors imported during the First World War. It was a light tractor with a horizontal twin-cylinder engine run on paraffin fuel. This photograph of a Titan being used for road haulage dates from about 1920.* [35/6282]

99 *The Titan was superseded in 1923 by a new International Harvester 10–20 tractor which, after the Fordson, was one of the leading types of tractor during the inter-war years. International adopted what was now becoming the conventional styling and equipped the tractor with a four-cylinder engine to run on petrol or paraffin. The principal novelty was the power take off, adopted by International as standard, one of the major advances in tractor technology, enabling power from the engine to be taken to drive trailed implements.* [5/1468]

100 *The Fordson model N, an example built in 1937. This tractor was novel in design when it was introduced in 1917 in being of frameless construction. The engine, gearbox and rear axle castings were made as strong cast iron units which when put together produced a rigid structure which did not need the support of a frame. The result was a tractor that was compact, light in weight, yet sturdy. It was also cheap to manufacture which helped Fordson maintain its policy of low prices, and that in turn contributed to the Fordson's dominance of the market. (A contemporary photograph of a Fordson in use in the 1930s can be seen on the back jacket, bottom.) [60/14024]*

Left **101** *Crawler tractors won considerable favour, especially on large farms, for their greater traction gave them an advantage for heavy work. Diesel power was applied more widely to tracklayers than to wheeled tractors in the 1930s. Both Caterpillar and Cletrac first marketed diesel versions of their tractors in 1931. Almost all the crawler tractors in use came from the United States: the Cletrac, for example, from Cleveland, Ohio. Here it is working with a cultivator in 1940.* [*K 22235*]

102 *The Ferguson-Brown tractor with a Ferguson system plough at a demonstration in Herefordshire in 1936. It had a Coventry-Climax engine of 18–20 hp, though production models after the first 500 were equipped with a 20 hp engine produced by David Brown. Despite the fact that it outperformed all rivals, sales remained low. Price was the main disadvantage, for at £224, the tractor cost nearly twice as much as the Fordson; added to which the farmer needed to buy the special Ferguson system implements, whereas with other tractors he could adapt horse-drawn implements in order to save money.* [*K 10521*]

DAVID BROWN
FOWLER
250C.

Top left **103** *In 1939 Harry Ferguson ended his agreement with David Brown, and transferred production of tractors using his hydraulic system to Ford in America. David Brown continued as a tractor manufacturer, and introduced the VAK1 in 1939, here demonstrating its work ploughing in 1940. It incorporated hydraulic lift and three point linkage, but not the complete hydraulic control covered by Ferguson's patents. Low-pressure pneumatic tyres had been available for tractors since 1932. Although that eased the difficulties of taking tractors onto public roads, pneumatic tyres gave poorer adhesion than steel wheels, and lower drawbar power. Consequently many farmers preferred steel wheels for field work, and tractors continued to be made with them into the 1940s. [35/10413]*

104 *The Fowler Gyrotiller. This machine was a rotary cultivator with two sets of tines rotated in opposite directions able to stir the soil to a depth of 15–20 inches without bringing the subsoil to the surface. Additional implements, principally the ridging bodies seen on this machine, could be attached. Power was provided by a diesel engine of 170 hp, with smaller versions between 30 and 80 hp also available. Originally devised for work on Cuban sugar plantations, Fowler introduced it into Britain in 1932. Contractors operated the machines regularly during the 1930s, but despite being generally well-received, and accorded approval in Claude Culpin's textbook on farm machinery, the Gyrotiller failed to establish itself permanently on the agricultural scene. [35/17040]*

Below **105** *The Wyles motor plough, manufactured by John Fowler & Co of Leeds, was one of a number of self-propelled ploughs produced during the years immediately before and during the First World War. Alfred Wyles had patented his motorized plough in 1911, his design being for a single furrow plough steered by a man walking behind. Subsequent developments by Fowler produced larger versions, the largest a four-furrow plough powered by a 45–50 hp engine, and with the driver seated. Motor ploughs never gained much popularity, however, because they were lacking in versatility compared with the tractor or the horse. This is one of the two-cylinder Wyles motors of 1913, equipped with a two-furrow plough. [35/8620]*

106 *More successful were the light two-wheeled market garden tractors of the 1930s. They were fitted with a tool bar to which could be attached row crop cultivators, harrows or a single furrow plough. The Plowtrac, manufactured in America, was introduced in 1939 by the British Motor Boat Manufacturing Company, which later went on to make similar machines in this country. The Rowtrac and the British Anzani 'Iron Horse' were among the other market garden tractors on the market at this time. [K 21599]*

Above right **107** *The Ransome MG2 market garden tractor was a compact crawler tractor powered by a 6 hp single cylinder Sturmey Archer engine, a type previously used by Ransomes to power lawnmowers. The tractor was introduced in 1936 and offered small farmers greater versatility than the two-wheeled tractor. It was extremely manoevrable, the tracks could be adjusted to suit different widths of rows, and a wide range of ploughs, cultivators and other implements could be used with it. It is shown in this photograph with a 'Demon' orchard sprayer made by W. G. Cooper & Sons of Wisbech. [35/26324]*

Right **108** *The 'Pioneer' engine by E. H. Bentall of Maldon was first shown at the Smithfield Show in 1913. It was a small four-stroke petrol engine, available as $1\frac{1}{2}$ and $2\frac{1}{2}$ hp. It cost £16 10s. [35/17975]*

BENTALL'S NEW "PIONEER" ENGINE.

See Stand No. 248, Smithfield Show, London, December 7, 8, 9, 10 & 11.

(Formerly known as the "Southwell").

Specially suited for

CHAFF CUTTING,
GRINDING,
PULPING,
SAWING,
PUMPING,
DAIRIES,
ENGINEERS,
BLACKSMITHS,
WHEELWRIGHTS,
&c.

Guaranteed Powers:

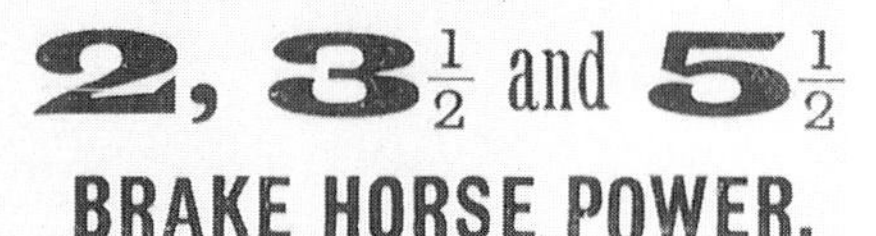

BRAKE HORSE POWER.

ANY SIZE MADE PORTABLE IF DESIRED.

Extremely Simple and Durable.

Write for further particulars and terms to

Above **109** *An advertisement for the petrol engines manufactured by R. A. Lister & Co of Dursley, Gloucestershire, 1908. [35/11253]*

Left **110** *The Ransome 'Wizard' paraffin oil engine, a vertical two-stroke engine, available in 3½ or 4 brake horsepower versions. It was introduced in 1921. [35/17980]*

Top right **111** *A 'Wizard' engine installed in a barn driving a Bamford No. 2 grinding mill and root cutter, 1923. [35/26320]*

Right **112** *The first generation of tractor ploughs required a ploughman riding at the back to steer, control the depth at which the plough was working and to lift the shares clear of the ground at the end of the furrow. They were known as steerage ploughs, and the ploughs controlled by the tractor driver, which became generally available during the 1920s, were self-lift ploughs. Ransome's RYLT plough, introduced in 1914, was a steerage plough available as a three- or four-furrow implement, and is seen here drawn by an International 'Mogul' tractor in 1915. [35/18510]*

C.6646
I.H.C. MOTOR TRACTOR
FURROW

10479.

Top left **113** *Ransome No. 9 RSLM plough illustrates the design of tractor ploughs that had become established by the 1930s. The rigid steel frame, of roughly triangular plan, was carried by three wheels, two forward, and a second furrow wheel at the back. Levers and screw mechanisms controlled the depth of ploughing, and a self lift ratchet operated by a lever that could be reached by the driver rendered the ploughman riding at the rear unnecessary. With the extra power of the tractor, multi-furrow ploughs became more usual. Disc coulters, rarely seen on British ploughs of the nineteenth century, were now a common option.* [35/18934]

Above **115** *An International Harvester 'Trac Tractor' crawler tractor with a three-furrow self-lift plough and four-furrow seam press at work in Berkshire, late 1930s.* [5/1405]

Far left **114** *Cockshutt ploughs entered the British market in 1933 as the result of an agreement whereby R. A. Lister were to sell the Canadian ploughs in this country, while Cockshutt would distribute Lister cream separators in Canada. Cockshutt offered a small range of strong, steel, self-lift ploughs, including their No. 6 plough convertible from two to three furrows, which won a fair following in Britain. Here, one of their ploughs is seen at a demonstration, with a Fordson tractor, in 1936.* [K 13191]

116 *The unchanging nature of some implements is illustrated in this photograph of a Cambridge roller in the Cotswolds during the 1930s. The Cambridge was of the clod-crushing type, a series of iron discs free to turn on an axle, and had been invented in 1844 by W. A. Cambridge. It continued to be made into this period in both horse-drawn and tractor-drawn versions.* [5/1547]

Top right **117** *A Ransome 'Dauntless' rigid tine cultivator, 1938. Tractor cultivators were constructed with a more rigid steel frame than was needed for the horse-drawn implements. There were levers similar to those on the tractor ploughs to control the depth of working and to operate the self-lift mechanism.* [K 19754]

Right **118** *One effect of tractor power is shown in this photograph. Hornsby's No. 7 seed drill had been introduced before the First World War as a light weight horse drill, intended to be drawn by two horses instead of three. Force-feed became more common in the twentieth century because it could sow the seed more evenly, and it was cheaper to manufacture.* [35/20246]

HORNSBY
FORCE FEED
HORNSBY
FORCE FEED
FORCE FEED

M H
D 2

Top left **119** *Massey-Harris gained a substantial share of the British market for seed drills during this time. Their drills used force-feed and disc coulters in preference to the Suffolk shoe-pattern. This is one of their combined seed and manure drills photographed in 1936. [K 10113]*

Left **120** *The disc harrow originated in the United States: the first patents for this type of implement were taken out as early as the 1840s. It was not until the end of the nineteenth century that it was introduced into this country, and because of its heavy draught it did not come into its own until the advent of tractor power. This wartime photograph shows a disc harrow being used to prepare a seed bed, hauled by a Caterpillar D2 diesel tracklaying tractor, a popular model from its introduction in 1938. [5/1920]*

121 *The spring-tined harrow was another American invention, patented in 1869. It was really a light cultivator with backward-curving tines so tensioned by the frame that they would spring back into position after encountering any obstruction in the soil. These harrows were first shown in Britain at the Royal Show of 1898. The one in this photograph was made by Massey-Harris in the 1930s. [K 13189]*

INTERNATIONAL

Top left **122** *Wilder's pitch-pole harrow, one of a number of harrows designed for work with tractors. Two sets of teeth were set within a frame. As the first set became clogged with rubbish, a pull on a trip cord would bring the second set into play, while the first was cleaned by the motion of the implement. [K 4817]*

124 *Ferguson-Brown tractor and stubble cultivator at work, 1936. [35/10425]*

Left **123** *Bamfords' self cleaning grass harrow, another harrow designed to clear the tines while it was in motion. The teeth moved up and down between stripper bars to clean off the rubbish. The harrow is being demonstrated in 1936, hauled by an International Harvester 10–20 tractor, and followed by an interested group of onlookers. [K 13185]*

125 *A four-furrow steerage row-crop cultivator and Fordson tractor hoeing sugar beet, 1943. Row-crop cultivators were of rigid construction, similar to field cultivators, and many were designed to be mounted on the power take off to control the depth and to lift the implement at the end of the furrows. Fordson, however, were slow to adopt power take off, hence the use of the steerage implement here. Fordson did, however, produce row-crop versions of their tractors with extra narrow wheels, and the body mounted high to allow tools to be carried underneath. The row-crop tractor was an innovation of International Harvester's in the United States, and was taken up slowly in this country during the 1930s. [35/26383]*

Top right **126** *A 'Robot' mechanical transplanting machine at work. With vegetables being more commonly grown as field crops there was a need for reducing the labour of setting out young plants. This machine could set out up to 10,000 plants an hour of such crops as sprouts, cauliflowers, or strawberries. The workers on the machine placed the plants into pairs of fingers on an endless chain which carried them down to the ground releasing them in an upright position. The inward slanting wheels compacted soil around the plants. The tractor, an International Harvester 'Farmall', demonstrates how the high-mounted body of the row crop tractor left clearance for cultivators to be mounted underneath. [K 23527]*

Right **127** *Potato ridger and fertilizer drill by John Wallace & Sons, Glasgow, 1939. It was mounted on the power take off which allowed precise control of the distribution of the fertilizer. It is hauled by an Oliver 70 row crop tractor. [35/26387]*

WALLACE

SPREADER
SIZE No. 2 3
C426

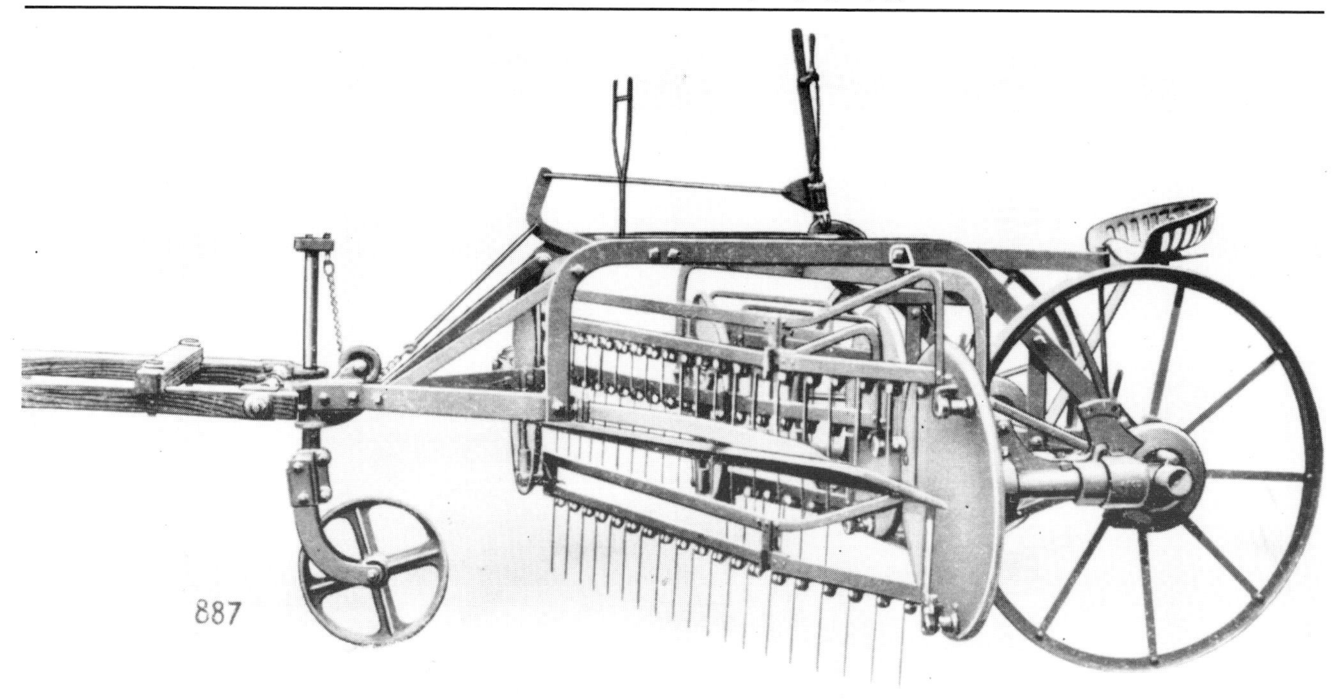

The Combined Machine set for Raking.

Top left **128** *The 'Success' manure distributor manufactured by the Paris Plow Company of Canada, attracted attention when it was first shown in this country in 1907–8 for its departure from the conventional designs which followed the structure of seed drills or of water carts. Instead, this was described by the manufacturers as being in the form of a farm wagon, with a mechanically operated revolving drum in place of the tail board. A movable apron shifted the manure backwards down the wagon. It was also intended mainly for farmyard manure rather than the artificial fertilizers distributed by the barrow-pattern implements. This type of machine continued to be available into the 1950s, made by Massey-Harris. [35/8021]*

Left **129** *The Wilder 'Cutlift' grass harvester was advertised, when it was introduced in 1934, as a 'grass elevator combine'. It consisted of a mower with an elevator and trailer. It could cut short and succulent grass, not the long hay only, and was taken up mainly by farmers who cropped grass for drying or for silage. [K 23741]*

130 *The side delivery rake was introduced in the early 1900s. It was a combined hay maker and collector. It gathered the hay from two swaths and delivered it to the side in a continuous windrow where it could dry before being collected for stacking. The side delivery rake could also be adapted for use as a swath turner. This example of the implement is by Nicholson of Newark. [35/26397]*

Top left **131** *Development in the 1920s and 1930s produced machines that could be used as hay tedders, swath turners and side delivery rakes. This photograph shows a Massey-Harris Blackstone machine being demonstrated in 1937. [K 15023]*

Left **132** *The hay sweep was an implement for gathering large loads of hay and sweeping them across the field to the stack. It was invented as a horse-powered implement, having a horse at each side and a steersman riding behind. It was redesigned to be mounted on the front of a tractor, and the photograph shows a farmers' economy measure, often employed in the 1930s, of using old cars or lorries to drive the sweep. [K 14922]*

133 *Appearing at the side of the previous photograph was the overshot stacker, often used in conjunction with the hay sweep since its cradle was of a size that conveniently took the load borne by the sweep. This stacker was a sort of catapult that swung the hay up from the ground and dropped it on to the stack. The lifting gear was operated by a winch, and that was best driven by a tractor. [K 14908]*

134 *The hay loader was an invaluable labour-saving implement for gathering hay. It was hitched on to a wagon and picked up hay from the swath or windrow and delivered it on to the back of the wagon as it moved along. The loader had a row of forks along the bottom oscillated by a drive from the main land wheels and these gathered the hay on to the conveyor. The conveyor in turn was a series of forks operated by chain drive. Knapp's 'Monarch' hay loader of the early 1930s offered improved adjustable forks, which, the makers claimed, would gather every last bit of the hay crop. [35/26007]*

Above **135** *Stacking the hay, showing the conveyor elevator, and the continued use of the horse gear to drive it.* [35/17311]

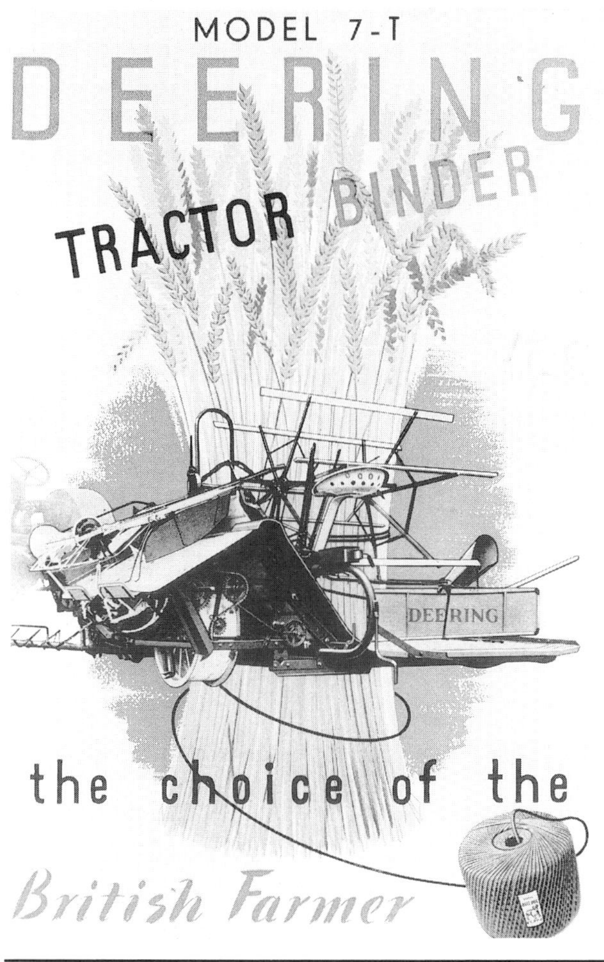

136 *The catalogue for Deering binders (part of International Harvester), 1938. Binders intended for tractor haulage were built with a more rigid structure to withstand the greater stresses resulting from the higher speed of travel across the field. They were also built to operate from the power take off on the tractor which meant that the cutting and binding mechanisms could continue at a constant speed independent of any variations in the progress over the ground. A third major change was the fact that the North American manufacturers, Massey-Harris and International Harvester were gaining a larger share of the British market for harvesting machinery.* [35/26389]

137 *A McCormick tractor power binder driven by an International Harvester 10–20 tractor cutting oats on the Berkshire Downs in 1936. Behind, the horses are pulling an Albion binder.* [5/1147]

A much appreciated feature of the Massey-Harris Steel Thresher is the ease with which adjustments can be made. Adjustments are quickly made from the outside of the machine, saving costly delays. The machine pictured above is equipped with the clover attachment.

SPECIFICATIONS OF MASSEY-HARRIS STEEL THRESHERS
22 x 36 in.; 24 x 44 in.; 28 x 48 in.

	No. 1 Steel	No. 1B Steel	No. 2B Steel
Width of Cylinder	22"	24"	28"
Width Rear of Machine	36"	44"	48"
Number of Bars in Cylinder	12	12	12
Number of Spikes in Cylinder	60	66	78
Number of Bands on Cylinder	3	3	3
Diameter of cylinder, including spikes	21⅜"	21⅜"	21⅝"
Speed of cylinder—R.P.M.	1140	1140	1140
Size of cylinder shaft	2" dia.	2" dia.	2" dia.
Length of cylinder shaft bearings	2⅞" Hyatt Roller Bearings on all sizes.		
Length of grate surface	20"	20"	20"
Front concave adjustment	Yes	Yes	Yes
Rear concave adjustment	Yes	Yes	Yes
Diameter of main drive pulley	8"	8"	8"
Face of main drive pulley	9"	9"	9"
Number of beaters	1	1	1
Length of straw racks	10' 6½"	10' 6½"	10' 6½"
Length of grain bottom and chaffer	12' 0⅞"	12' 0⅞"	12' 0⅞"
Rack surface in sq. ft.	30.9	37.77	41.36

	No. 1 Steel	No. 1B Steel	No. 2B Steel
Chaffer surface in sq. ft.	11.85	14.48	15.8
Riddle surface in sq. ft.	12.66	13.25	14.16
Length of Separator with W.S. and S.F. carrier, folded	23' 9½"	23' 9½"	23' 9½"
Length of Separator with W.S. and S.F. carrier, extended	44' 9"	44' 9"	44' 9"
Height of machine at Feed Table	4' 10"	4' 10"	4' 10"
Height of machine at rear	7' 6"	7' 6"	7' 6"
Height of front wheels	30"	30"	30"
Height of rear wheels	34"	34"	34"
Width of front wheels	5"	6"	6"
Width of rear wheels	5"	6"	6"
Track of front wheels C. to C.	4' 8½"	4' 8½"	5' 0½"
Track of rear wheels C. to C.	5' 8⅝"	6' 1⅝"	6' 8⅝"
Average capacity wheat	60	100	125
Average capacity oats	100	150	175
Belt H.P. required with S.F. and W.S.	10 to 25	15 to 30	30 to 40
Weight fully equipped less Grain Weigher	6570	6800	7070

REGULAR EQUIPMENT—MASSEY-HARRIS STEEL THRESHERS

No. 1—22 x 36 in. Steel Thresher, Wind Stacker, Massey-Harris Feeder, 9-ft. Carrier, No. 1 Perfection Register.
No. 1B—24 x 44 in. Steel Thresher, Wind Stacker, Massey-Harris Feeder, 9-ft. Carrier, No. 1 Perfection Register.
No. 2B—28 x 48 in. Steel Thresher, Wind Stacker, Massey-Harris Feeder, 9-ft. Carrier, No. 1 Perfection Register.
SPECIAL ORDER EXTRAS: 14-ft. Carrier for 24 in. or 28 in. Massey-Harris Feeder, Clover Attachment, Chaff Blower, Hinged Stacker Pipe. Various types of Hart Registers or Loaders. Also, Hart or Garden City Feeders. Straw Bruiser.

Left **138** *Massey-Harris and International Harvester introduced all-steel threshing machines to the British market. The steel body was reckoned to make the machine more resilient to the stresses of work than the traditional wooden frame. Internally there was some redesign, with stripping knives ahead of the drum, and the series of shakers and fans arranged such that the whole machine was longer and thinner. They were also designed for bulk handling, with pipes delivering the threshed corn, and a 'wind stacker' pipe (rarely used in British conditions). Massey-Harris catalogue of 1938. [35/26388]*

139 *An International Harvester steel threshing machine at work, 1939. The extended self-feeding mechanism was a feature of these machines. The corn delivery pipe is filling sacks in the conventional British way, rather than bulk trailers, and the straw is being fed into a straw-baling press with a conveyor up to the lorry. [K 19574]*

UD4743

Top left **140** *Straw-baling press with overhead feed, similar to the one working with the steel thresher in the previous photograph. This one was manufactured by the German Claas company, and introduced into Britain in 1934. [K 8777]*

Left **141** *One of the first combine harvesters to be produced by a British firm was the Clayton, being demonstrated here in 1932. In common with all combine harvesters of this time, the machine was hauled by a tractor. The cutter was set well to the side of the thresher. The threshed grain was delivered straight from the machine to be filled into sacks, and this required a number of men on hand to maintain the supply of sacks. [K 2233]*

142 *A Sunshine 'auto-header' combine harvester at work, 1940. Sunshine grain harvesters were manufactured in Australia by a subsidiary company of Massey-Harris. During the late 1930s and into the Second World War Sunshine binders and combine harvesters were imported in considerable numbers. [35/6266]*

MASSEY-
HARRIS
MASSEY-HARRIS

Top left **143** *Self-propelled combine harvesters were brought to Britain at the beginning of the Second World War. Massey-Harris introduced two in 1941, the 21, a medium-sized machine with 12 feet width of cut, and the larger 20, shown here, which could cut a width of 16 feet. Both were tanker combines, a type which arrived on the market in the early 1930s and which eliminated the need for the sack-filling gang. [35/25394]*

145 *A Case pick-up baler at work in 1938. The tractor is an International Harvester 'Farmall' F20 row-crop tractor introduced in 1932. Pick up balers reached this country from the United States in the early 1930s, used initially for baling hay, and subsequently for baling the straw left by the combine harvesters. Even by the end of the Second World War, however, they were not particularly common implements. [K 18596]*

Left **144** *Massey-Harris combine harvester and British Wallis tractor on a demonstration, 1929. [35/17293]*

Top left **146** *A John Deere potato digger, 1941. American potato diggers, which used a broad-share shovel to dig up the row of potatoes and deposit them on a riddle elevator to shake off the earth, were introduced into Britain late in the 1930s. They relied on row crop tractors to provide the power for the riddle conveyor, and that retarded their acceptance, since many potato growers still used horse power only. [35/7434]*

Left **147** *The expansion of the acreage of sugar-beet during the 1920s and 1930s was achieved with little mechanization. Only in the 1940s were effective beet-harvesting machines beginning to be developed. This is a Catchpole harvester of 1944. [35/26384]*

148 *Mechanization came late to dairying. The milking machine had been invented as early as 1862, and developed mainly in America during the following 15 years. By the 1890s British, American and European manufacturers had machines on the market. But farmers were slow to adopt milking by machine. It was not until the Second World War that the numbers of machines in use increased sharply: an increase of 64 per cent in those five years brought the proportion of the national dairy herd milked by machine up to about 50 per cent. The most common type of milking machine in use in this period was the portable pail shown in use on a Cambridgeshire farm in 1937. [K 10370]*

The Great Success of 1909, the New C Model Lister will beat all-comers in 1910.
LISTER
CREAM
SEPARATOR
"MADE THROUGHOUT IN BRITAIN"
Simplicity Itself
Agents now wanted for a few unrepresented districts;
Apply without delay to R·A·LISTER & Co L^{TD} DURSLEY.

Left **149** *An advertisement for a Lister cream separator from 1910. Cream separators, developed in Germany and Scandinavia, were introduced to Britain in the late 1870s. [35/15249]*

150 *The production line of Ferguson-Brown tractors in 1937. [35/10341]*

Index

Entries printed in bold refer to illustrations